BUYER'S GUIDE TO
MODEMS & COMMUNICATIONS SOFTWARE

Also by the Authors from TAB Books, Inc.

No. 1872 *Buyer's Guide to Computer Printers*

The authors wish to thank David Wilkes and the people at Racal-Vadic for their cooperation in technical material presented in this book.

No. 1882
$19.95

BUYER'S GUIDE TO MODEMS & COMMUNICATIONS SOFTWARE

TERRY C. SILVERIA,
SANJIVA K. NATH,
& EDWARD HOGG

TAB BOOKS Inc.
BLUE RIDGE SUMMIT, PA. 17214

FIRST EDITION

FIRST PRINTING

Library of Congress Cataloging in Publication Data

Silveria, Terry.
Buyer's guide to modems and communications software.

Includes index.
1. Modems—Purchasing. 2. Computer programs—Purchasing. I. Nath, Sanjiva K. II. Hogg, Edward. III. Title.
TK5103.S52 1985 001.64'42 84-23512
ISBN 0-8306-0882-6
ISBN 0-8306-1882-1 (pbk.)

Contents

Introduction vii

1 **Applications** 1
BRS After Dark—CompuServe—DIALOG—The Dow Jones News/Retrieval Service—The SOURCE—Electronic Mail—Home Banking—Networking—Electronic Catalogs—Electronic Extension Classes

2 **Digits and Data** 6
Counting Digits—Semiconductors—Combinations—Binary Numbering—Microprocessors—Programming—Summary

3 **The Modem** 12
Serial Transmission—Intelligent Modems—Modem Arrangements—Communications Software

4 **Modem Standards and RS-232C** 18
Bell Standards—Vadic 3400—CCITT Standards—The RS-232C Interface—Other Interface Standards—Beyond the Standards

5 **The Phone Connection** 25

6 **Communications Software** 28
Control of the Modem—Program Drive—Main Features

7 **Selecting a Modem and Communications Software** 33
Selecting a Modem—Selecting Communications Software

8 **Modem Reviews** 37

Anchor Automation, Inc—Apple Computer, Inc.—Business Computer Corporation (Bizcomp)—Cermetek Microelectronics Inc.—Commodore Business Machines, Inc.—Communications and Peripherals Marketing Corporation—Computer Communications Specialists, Inc.—Concord Data Systems—Datasolvers, Inc.—Gandalf Data, Inc.—General Datacomm, Inc.—Hayes Microcomputer Products, Inc.—Human Engineered Software (HESWARE)—IDE Associates, Inc.—Lexicon Corporation of Miami—MFJ Enterprises, Inc.—Microbits Peripheral Products, Inc.—Morrow, Inc.—Multi Tech Systems, Inc.—NEC America, Inc.—Novation Inc.—Penril Datacomm—Prometheus Products, Inc.—Racal-Vadic—Rixon, Inc.—Tek-Com Corporation—Televideo Systems, Inc.—The Microperipheral Corporation—TNW Corporation—Transend Corporation—Universal Data Systems—U.S. Robotics, Inc.—Ven-Tel, Inc.—Visionary Electronics

9 **Communications Software Reviews** 129

Alpha Software Connection—Apple Computer, Inc.—Bi-Tech Enterprises, Inc.—Communications Research Group, Inc.—Computer: Applications, Inc.—Digital Marketing—Direct Aid, Inc.—Dynamic Microprocessor Associates, Inc.—Execs Software—Hawkeye Graphix—Hayes Microcomputer Products, Inc.—IDE Associates, Inc.—Lindbergh Systems, Inc.—Mark of the Unicorn—Microstuf, Inc.—MicroSystems Software—Microtechnic Solutions Inc.—Midwest Micro Inc.—Morrow, Inc.—Mycroft Labs, Inc.—New Era Technologies, Inc.—Sanyo Business Systems—Software Publishing Corporation—Southeastern Software—Supersoft Inc.—Tau Systems Corporation—Transend Corporation—United Software Industries—U.S. Digital Corporation—U.S. Robotics, Inc.—Vector Graphics, Inc.—Windmill Software, Inc.—Zoom Telephonics, Inc.

Appendix A Product List 163

Appendix B Vendors Addresses and Phone Numbers 167

Glossary 172

Modem Comparison Chart 177

Index 182

Introduction

Computers, which are designed primarily to process information, can transfer information as well. Through the process of telecommunications, information generated in one computer can be transferred great distances using ordinary phone line. Through telecommunications, the resources and data processing capabilities of the most advanced computers become available to owners of inexpensive home systems.

Regardless of the make or model of computer, there are two essential ingredients in adapting a home system for telecommunications: a modem and communications software. The *modem* is the device that connects the computer to the telephone system. It turns signals used by the computer into signals that can transfer across the phone lines. Like all other equipment connecting to computers, modems come in a variety of types and models. They include long lists of specifications that may or may not be of importance for a specific application.

All computers require programs to direct their activities. A *communications program* commits a computer to the activity of telecommunications. It regulates and controls the transfer of data that occurs through a modem. The features and the method of implementing them differ between programs, however. Some communications programs are designed for specific applications requiring specific features; others are designed with an emphasis on simplicity and ease of use. Selecting just the right program for your needs is often a considerable challenge.

We have written this book for those considering the purchase of a modem or communications software for any personal computer. We doubt that many readers will read this book from cover to cover. It is meant to serve as a reference for those attempting to select from numerous products the one best suited to their needs. The early chapters provide a background on modems and communications software which should help those completely new to the field of telecommunications. You may find this material useful in understanding terms used in the reviews. Additionally, the glossary at the end of the book provides a quick reference for all the technical terms.

The emphasis of the book is on providing information on the product themselves. For each of the products reviewed, we have provided the information necessary to make an informed selection. Because the suitability of a product depends on the particular application, we have not attempted critical reviews. Instead, we have reported, as objectively as possible, the various features found in each product. We have made a point of including addresses and phone numbers of vendors of all products; the business of modems and communications software is becoming quite competitive, and vendors are always happy to provide additional information to potential customers.

1

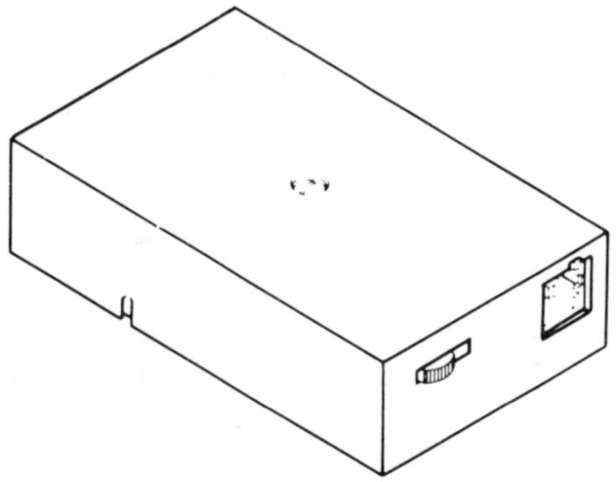

Applications

Through telecommunications, applications that are normally associated with sophisticated mainframe computers are accessible to the personal computer. Numerous commercial information systems collect, process, and save information on various topics. These commercial systems make this information available for a fee. To be profitable, these need a large customer base. The growing interest in telecommunications promises to expand that potential customer base and increase the number of commercial systems available. The following pages provide a brief description of some of the more popular commercial systems and applications possible through telecommunications.

BRS AFTER DARK

BRS After Dark is an off-hours, general-public version of Bibliographic Retrieval Service, a research tool created in 1976 for businesses, universities, and libraries. At a lower cost, this trimmed-down but still high-quality version makes available the most popular 25 data bases from the parent service's 75. It provides references and abstracts from magazines, journals, and industry and government publications about Science and Medicine, Business and Finance, Education, Social Science and Humanities, and Energy and Environment.

You access the data bases of BRS After Dark through progressive menu selections. Your own keywords then zero in on a subject. Proprietary software helps you to tailor the screening of citations and abstracts. In this way, an intimidating mass of material quickly refines down to the pertinent essence of a research project. Auxiliary services include Newsletter (on-line), Software (for downloading), Electronic Mail (for addressed access), and Swap Shop (a bulletin board).

Local access times are: 6:00 P.M. to midnight Monday through Friday, and all day Saturday and Sunday. Costs are: $50.00 advance and $6.00/hour to $15.00/hour usage ($12.00/month minimum). Address: *Bibliographic Retrieval Services, 1200 Route 7, Latham, NY 12110.* Information: *(800) 833-4707.*

COMPUSERVE

CompuServe is a highly varied, dynamic amalgamation of services, ideas, opinions, and facts. It fosters the curiosity and practical needs of active people of all ages and occupations. It caters to general concerns (education, shopping, travel, and entertainment) and to special interests (business, finance, medicine, and bulletin boards for computerists, professionals, and enthusiasts of all kinds).

The following categories are only a sample of CompuServe's wide-ranging labyrinth of tantalizing resources:

Communications. Keyboard version of CB chit-chatting; electronic mailboxes for sending and receiving messages; feedback forum; general bulletin board for notices and personal ads; directory of special-interest users.

Business and Financial Information. Wire services, reports, newsletters, stock performance, commodity information, and investment-analysis tools.

Shopping and Banking. Comp-U-Store allows comparison and informational shopping and ordering. Product Catalogs; shopping tip sheets; monitoring and handling bank accounts.

Education. Advice on college selection and preparation; newsletter for deaf students; games, quizzes, puzzles, and contests; *The World Book Encyclopedia.*

Entertainment and Travel. Art printouts; wine selection; travel services; airline schedules; flight planning; vacation homes exchange; bargains and events.

Science and Medicine. On-line forum on health-care services; aviation safety; Food and Drug Administration news service; information exchange on animal care.

Computer Services. Download purchased or free software; use mainframe software; computer company newsletters; meeting ground for users and hackers; 128K free storage.

CompuServe has a one-time joining charge of about $40.00, usually purchased through a computer store, which includes 5 free hours of use. Hourly rates for 300-baud transmission are: $22.00 daytime and $5.00 nighttime, billed directly or by VISA or MasterCard.

Address: *CompuServe, Inc., 5000 Arlington Centre, Box 20212, Columbus, OH 43220.* Customer service: *(614) 457-8650* in Ohio; *(800) 848-8990* in other states and contiguous countries.

DIALOG

DIALOG is a huge, in-depth bibliographic service for serious research. Run by the Lockhead Corporation, it developed from the need of NASA researchers for rapid access to timely information in many fields. Since becoming available to the public in 1972, it has attracted over 500,000 users in more than 50 countries.

The full DIALOG service has more than 75 million records in nearly 200 data bases, covering Business/Economics, Chemistry, Agriculture and Nutrition, Medicine and Biosciences, Energy and Environment, Science and Technology, Patents, Law and Government, Current Affairs, and Directories. The bases provide on-line bibliographic data, citations, and abstracts. You can also order full-text copies of nearly every item listed, and a Selective Dissemination of Information service can keep you aware of updates.

The standard DIALOG service operates 22 hours a day, Monday through Saturday. Usage rates range from $10.00/hour to $285.00/hour, depending on the data base. Text printouts cost extra, and you will need a $40.00 DIALOG manual.

Since 1982 DIALOG also has had the KNOWLEDGE INDEX, a low-cost specialty service for the home computerists. It makes available, outside of normal business hours, some of the most popular bases. These are: ABI/INFORM (over 500 publications), AGRICOLA, Computers and Electronics, Engineering Literature Index (of about 3500 journals), ERIC, GPO Publications Reference File, INSPEC, International Software Index, Magazine Index, MEDLINE, Microcomputer Index, National Newspaper Index, NEWSEARCH, PsychINFO (covers over 900 periodicals, 1500 books, worldwide technical reports and monographs), and Standard & Poor's News Online.

KNOWLEDGE INDEX hours (local time) are:

Monday through Friday 6:00 P.M. to 5:00 A.M.; Saturday 8:00 A.M. to midnight; Sunday 3:00 P.M. to 5:00 A.M. Sign-up costs $35.00, which includes a manual and 2 free hours of use. Connect time costs $24.00/hour.

Address: *DIALOG Information Services, Inc., Marketing Department, 3460 Hillview Avenue, Palo Alto, CA 94304*. KNOWLEDGE INDEX Customer service: *(800) 353-0458* in Arizona, *(800) 528-0470* in Alaska and Hawaii *(800) 528-6050* in other states and Canada.

THE DOW JONES NEWS/RETRIEVAL SERVICE

At this time, The Dow Jones News/Retrieval Service (DJNS) mostly benefits active investors who need comprehensive, historical and timely facts, news, and analysis about the world of business and finance. Run by the publishers of the *Wall Street Journal* and *Barron's*, the service makes available the past and current coverage of those and other publications, along with *World Report* from United Press International. It gives current and historical quotes from financial trading markets, corporate earnings estimates, digests of company disclosure data, and various economics surveys and commentaries.

DJNS also caters to less-specialized interests. It has movie reviews, sports news stories and scores, domestic and international weather reports, discount shopping through Comp-U-Serve, and an on-line version of the 21-volume *Academic American Encyclopedia,* searchable by article, title, or subject heading. DJNS also has a physical link to BRS (see BRS After Dark) and a powerful Free Text Search.

DJNS has a one-time-only $50.00 start-up fee, sometimes avoidable through various promotional deals. Prime-time 300-baud access rates range from $9.00/hour to $72.00/hour, depending on the data base. Non-prime-time (6:01 P.M. local to 4:00 A.M. EST, and all day weekends and holidays) access is substantially but variably cheaper. It also has at least two special plans: Executive Membership and Blue Chip membership. All billing is direct; no credit cards.

All subscribers receive *Fact Finder,* a manual which describes all the ins and outs of network sign-on procedures and codes for reaching into different data bases. DJNS also distributes a quarterly support publication, *Dowline,* with information about new data bases, new features, and software for accessing DJNS information.

Address: *Dow Jones News/Retrieval Service, P.O. Box 300, Princeton, NJ 08540*. Customer service: *(609) 452-1511* in New Jersey *(800) 257-5114* in other states and Canada.

THE SOURCE

The SOURCE describes itself as an information utility, America's first. It makes available continually updated information of varying depth but of considerable range. Usefulness to busy, multi-interest people seems to be its emphasis, although no one would likely fully explore all of its resources.

Many of its 1,000 or more files are full-sized data bases. It provides new wire services (UPI, finance, weather, sports, syndicated columns), abstracts from major magazines and business journals, and financial and political newsletters. You can access this information yourself, or obtain customized research for specific projects, and obtain monthly reports to keep track of specified topics.

Some of the files support interactive services, such as electronic mailboxes, voicegrams, mailgrams, bulletin boards, teleconferencing, discount shopping (Comp-U-Store), downloading free or bought software, and harnessing a mainframe computer. You can also barter and trade, search for a job, and solicit advice on medical or personal problems. The SOURCE even has an astrologer who will analyze your prospects.

Travel and entertainment get ample treatment. You can check on tourist amenities, consult airline schedules, and book tours, flights, cars, and rooms. Reviews cover movies and restaurants. You can play over 60 on-line games.

Games also feature in education services. You can have interactive quizzes and drills developed by the *Reader's Digest* Educational Division. They range from special programs for children through college level.

The SOURCE provides access to its services through a sequence of menus. The menus lead you through multiple layers of subdivisions until your choice narrows to a specific topic, citation, or whatever. After you become familiar with a particular data base, you can short-cut through to your destination with the system's standard commands.

The one-time sign-up fee is $100.00. For this you get a manual, a monthly newsletter, and a 4K personal storage space. Hourly rates range from $5.75 to $25.75, depending on baud rate and time of day. Other charges include: 25 cents for each sign-on, $1.00/month for account maintenance, and $9.00/month minimum usage fee.

Address: *Source Telecomputing Corporation, 1616 Anderson Road, McLean, VA 22102.* Customer support: *(800) 336-3300* in the continental United States *(703) 734-7540* in Virginia and outside the continental U.S.

ELECTRONIC MAIL

Communications vitalize all businesses: between salespersons and clients, between management and staff, between government and corporations; the list goes on. In the past, business communications occurred through spoken and written exchanges, either directly or by phone, wire, and mail. Recent developments in telecommunications have led to a communications hybrid: electronic mail.

Electronic mail works something like a telephone answering system, but uses written instead of spoken messages. The sender keyboards the message at a computer terminal and transmits it over the phone lines for deposit at an unattended terminal. On returning, the distant operator can, in a manner comparable to checking out a phone answering machine, retrieve the message by displaying it on the screen. The system can work between two personal computers, each leaving messages on the other. Commercial services allow the deposit of messages on a mainframe computer for retrieval only by the intended recipient.

Electronic mail has numerous advantages over traditional forms of communication: It is certainly faster than mailing printed matter, and it is usually cheaper than most phone calls. Phone conversations often consume much time inefficiently, whereas a busy executive can screen electronic-mail messages to exert some discretion over the time he spends communicating. Electronic mail well suits those whose work frequently takes them out of the office; attempting to reach a field worker by phone can often be quite frustrating. Some systems can address messages simultaneously to a number of recipients, such as those sent by a manager to a group of employees.

Electronic Mailing Service

Some firms currently provide electronic mailing services. A subscriber enters through a computer keyboard the text and an address and transmits them through the phone lines to a receiving center. The service prints the letter, seals it in an envelope, and delivers it to the addressee. One such service, through MCI, provides to many cities a 4-hour delivery time. While more expensive than a conventional letter, electronic mailing services avoid the expense and time for conventional labeling, posting, and mailing.

Electronic Bulletin Boards

Electronic bulletin boards, an interesting phenomenon of personal-computing telecommunications, are open-access electronic mail. Each bulletin board is, essentially, a test file at a particular phone location that allows many people, through personal computers and the phone lines, to add or read messages. It is an electronic version of the conventional bulletin board found in schools, businesses, and stores. Both kinds of boards have a similar wild variety of messages.

Typical bulletin boards are free, being operated by dedicated groups of computer enthusiasts. Phone numbers for local boards are often available through user groups or newspapers. The format for writing or reading messages varies with the bulletin board. Often, messages simply remain in chronological order. A special command from your terminal allows you to scroll through the messages.

Another command lets you leave a message. More sophisticated boards may let you address messages to individuals.

Some bulletin boards specialize in a specific activity. For example, followers of IBM computers may have their own board for leaving technical questions and answers. Buyers and sellers of used furniture, cars, or computers may come together through a board. You can even come together through dating boards: names and lists of personal qualities and interests, some quite explicit, line up electronically for all to read.

HOMEBANKING

Recently, a number of banks have introduced pilot programs for homebanking by computer. Subscribers can obtain a current account balance, find out which checks have cleared, transfer amounts between accounts, and, in some cases, pay bills. Identification numbers and a secret password ensure that no one else can probe your accounts.

A homebanking service does not allow for depositing or withdrawing money. Instead, it provides a convenient way to monitor your accounts and perform the more routine banking operations. Rather than wait until the end of the month to receive a check register and an account balance, subscribers can check their banking status immediately, at any time, day or night.

Some homebanking services allow bill paying: A subscriber can transfer funds from a checking account to a payee account set up for a specific company. For example, from your computer keyboard you can transfer funds from a checking account to the phone company to automatically make a monthly payment. You no longer need to make out a check, fill out the monthly statement, and mail it in. Through telecommunications, it all happens electronically.

NETWORKING

Some of the first commercial computer systems were designed around large mainframe computers. Individual terminals, disk storage devices, and printers connected directly to the main computer. The mainframe performed all the data processing and managed access to the various devices so that no two devices tried to use the same printer at the same time. An alternative to mainframe computers, that of linking together individual microcomputers in networks, is growing in popularity. *Networks* allow a group of microcomputers, each with its own data processing capability, to share resources, such as printers or disk storage devices. Special software, for managing access to various devices, avoids conflicts.

Networks are becoming quite popular in corporate environments. Through telecommunications, they often extend outside offices. For example, workers at home can dial up their office network, access files from a disk storage device, and have correspondence printed on an office printer. Often, the software necessary for accessing networks is more sophisticated than typical communications software. It may include some form of terminal emulation and some hardware adaptations.

ELECTRONIC CATALOGS

Some retail stores are experimenting with electronic catalogs. Holders of accounts at retail stores can access, through a computer, a data bank with descriptions of various products. Still at the computer, the shopper places an order. The merchandise arrives C.O.D. Generally, this process reduces much of the expense involved in retail sales, making possible discount savings on many types of products.

ELECTRONIC EXTENSION CLASSES

Through telecommunications, some universities are offering accredited classes, called *electronic extension classes*. The class material displays on the student's screen at the student's own pace. The student can direct questions to the instructor with the host computer. They are especially well-suited to students who have difficulty in commuting to a school but desire a more immediate exchange than correspondence classes. Electronic classes may also alleviate the overcrowding found in more conventional schools.

2

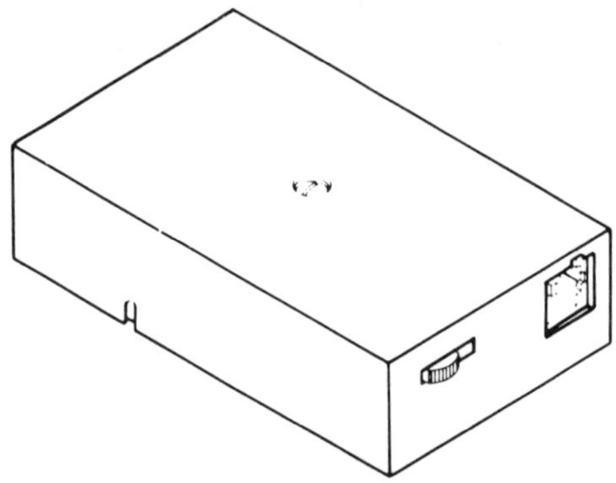

Digits and Data

This chapter looks at the general principles that underlie the more detailed later chapters. It examines the time-tested basics, those that will likely endure beyond the current crop of computer peripherals. In doing so, it should ease the confusion that product specifications often cause in nontechnical users. This brief account can, however, only touch on topics that really require a whole book. So, it mainly offers insights: Why digital? What is machine language? It also includes useful-to-know specifics: What is a bit? How big is a byte? Do bytes move in single file or side by side?

With luck, these topics might strike the reader as inherently interesting. More pertinent, however, and perhaps more reassuring, is the fact that the principles apply increasingly to more commonplace products. For example, recent automobiles, cameras, and electrical appliances will likely have some digital circuitry. Laser video discs and digital sound recordings both embody digitized information, the common currency of digital computers and printers. They all draw upon a family of devices which can store, reshape, and move around virtually any recordable variable, such as sound, pictures, words, and numbers.

Words and numbers illustrate most readily, perhaps, the digitizing of information, this chapter's primary focus. They form the raw, undigested data most immediate to the typical computer user, and they fit into the notions of language and counting inherent in computing. Of course, words and numbers represent concepts that a mere machine can in no way comprehend; yet the computer somehow converts those concept symbols into machine-manageable symbols. It also rearranges, files, and rewrites them at a distance with a swiftness unmatched by human hands.

COUNTING DIGITS

Computers, marvels that they are, contain nothing as marvelous as a human hand. They do, however, emulate a particular handlike function. In a certain sense, computers count on fingers of a sort, re-enacting the roots of conventional numbering. Their imaginary fingers come in pairs: the open and closed states of a switch. It has a myriad of these switch hands, clasping and unclasping, so to speak, in ever-varying exchanges. The term *handshaking* refers to a manner of handing over data from one digital device to another, and many hands make light work. So, however seemingly complex, computing amounts essentially to manual labor: a mass, incredibly rapid, checking-off of digits.

No accident accounts for why *digits* refer to both figures and fingers. It comes from the Latin for finger: *digitus*, and through the ages, our ten digits have formed an always handy computational device. Little wonder, then, that our ordinary numbering system uses tens. Named after the Latin for ten, *decem*, the now universal and seemingly obvious *decimal system* evolved over millennia, however. It was a collective invention, one of the world's greatest, spurred by everyday needs and curiosity about nature. It built upon occasional bright ideas, such as digit symbols, powers-of-ten positioning, the zero, the decimal point, and so on. These smart moves magnified progressively our manual and mental reach. They fostered science and technology, making possible today's prodigious number manipulators: digital computers.

At the most basic level, however, practical digital hardware cannot conveniently count in tens. Of course, digital devices may have readouts based on tens, as do digital watches, but their underlying technology does not. Primarily, this is because a tens base would considerably increase their complexity, unreliability, and cost. So, for this and other reasons, digital technology uses a simpler, but similarly powerful, system of counting based on twos: the *binary* system.

At first glance, numbering by twos seems a backward step, a restricting and unnecessarily tedious chore. It suits computers just fine, however. With utmost dependability, without rest or protest, the computers relentlessly, invisibly, and silently flip-flop their colonies of supertiny switches. Properly sequenced, those switches can work wonders.

SEMICONDUCTORS

Perhaps we need to qualify the use of the term *switch*. Computers do have switches in the usual sense of, say, an electric light switch. Such ordinary switches have movable contacts which can activate a circuit by physically joining two or more conductors. In fact, the computer keyboard has many momentary-contact switches. As we will discuss later, each keyboard key triggers its own circuit for issuing the standard pulse pattern for that symbol or command. The computer proper, however, contains mostly motion-free switches, which are the so-called *solid-state* devices which function as electrical gates. They are also called *semiconductors* because they contain normally resistant materials that become conductive on being electrified in a special way.

Semiconductors originated in the late 1940s as *transistors*, being used then in conventionally wired circuits to replace vacuum tubes. They evolved, however, into integrated parts of highly complex and microscopically small printed circuits—the so-called *integrated circuits*. These tiny pieces of silicon, commonly referred to as *chips*, are cheap to produce, having no moving parts to wear out or to waste energy, and make possible high operational speeds.

COMBINATIONS

The computer's real power lies in its capacity to juggle, at fantastic speeds, ways of combining opposites. Consider that a single two-state switch offers only two options: on and off. Two such switches together, however, have 4 variations: on/on, off/off, on/off, and off/on. Add another switch, and you get 8 possibilities: on/on/on, on/off/off, on/on/off, on/off/on, off/on/on, off/off/off, off/on/off, and off/off/on. Each additional switch multiplies the possibilities by two. For example, eight switches yield 256 possibilities. This number comes from multiplying together eight twos, mathematically known as two to the power of eight (2^8).

Clearly, the basic reasoning applies to combinations of any two choices: positive/negative, high/low, charged/uncharged, attract/repulse, clockwise /counterclockwise, and so on. The digital computer, in addition to on /off, employs several of these pair options because the computer does more than just compute; it does a lot of comparing—testing to see if programmed conditions are satisfied. It does so with the aid of a system of logic based on twos, *binary logic*, which was developed in the last century, long before applications were apparent or practical, but that gets into areas too far from the purposes of this book.

We are more concerned here with the end result of digital processing: digital output—the signals that pass between digital devices. Those signals are the moving counterparts of the static combinations of opposites just described. To grasp more readily the nature of these combinations, both static and dynamic, we need a system of identifying them. For this reason, we turn briefly now to the previously mentioned twos-based numbering system: binary numbering.

BINARY NUMBERING

Binary numbers have exactly the same structure as decimal numbers; the only difference is that one has a twos base and the other a tens. Both use *positional notation*, where a digit's relative position indicates its relative value. In decimal numbering, for example, each one-position shift to the left multiplies a digit's value by ten. Similarly, binary digits double in value for each shift to the left. Consequently, binary numbering uses only two digits: 0 and 1. The 0 always counts zero, regardless of position, but 1 has unitary value only in its primary position. Each move to the left doubles its value, as the following binary and decimal equivalents show:

$$\begin{aligned} 000001 &= 1 \times 1 = 1 \\ 000010 &= 1 \times 2 = 2 \\ 000100 &= 2 \times 2 = 4 \\ 001000 &= 4 \times 2 = 8 \\ 010000 &= 8 \times 2 = 16 \\ 100000 &= 16 \times 2 = 32 \end{aligned}$$

Intermediate values result from adding the components of a composite number. For example:

$$010101 = 16 + 4 + 1 = 21$$

Note that some of the zeros in these examples serve only to keep the number of binary digits constant. This is appropriate in a digital-computing context to maintain a fixed word size, as discussed later. Note also that binary numbering, as decimal numbering, can also accommodate fractions. The fractional part of a binary number also goes to the right of a point, called a *binary point*, which looks exactly the same as a decimal point. To the right of the binary point, the digit 1 takes the values half, quarter, and so on. Our principal interest here is with whole binary numbers, however, using them not to do arithmetic but to code information. Before we get to coding, we must touch on the significance of binary numbers to the computer's internal workings.

MICROPROCESSORS

You will recall that microcomputers contain a number of integrated circuits which are highly condensed networks of many semiconductor devices. These devices jostle around minute bundles of electrical energy at great speed. They interact those bundles with other bundles—adding, subtracting, and comparing—and briefly, here and there, hold them poised for action. In the poised state, the energy bundles exist as imbalances of electrical charge which have the tendency, or the *potential* (measured in volts), to restore equilibrium. They do so by discharging through a complex maze, as routed by traffic-controlling semiconductor gates. The charges do not actually criss-cross the maze intact. Rather, they trigger into existence other charges at other locations. They thereby set up intermittent and circuitous chain reactions, zigging and zagging to the beat of a high-frequency pacemaker: the computer's internal clock.

Clearly, groups of such charges can assume any of the two-state combinational possibilities noted earlier. Comparable groupings of any pair of symbols can represent such combinations. For example, the digit 1 could represent a charge, and the digit 0 could represent an absence of charge, or vice versa. The result: a binary number. By implication, the circuitry can move numbers around in the form of groups of electrical charges.

The integrated circuit which does the charge/number manipulating, or *processing* is called a *microprocessor*. It contains the *Central Processing Unit* (the CPU), which is the heart of the computer. This is the network of tracks, crossings, bridges, underpaths, sidings, and switch points that marshalls trains of electrical charges between semiconductor freight docks. These ghost trains have a fixed length, and some of their carriages remain empty. They all exit transformed.

A group of eight charges, which moves as a unit within or without a digital device, is usually called a *byte*. It also variously takes the names *word* or *character*. Each element of a byte (not necessarily an electrical charge) is called a *bit*, from BInary digiT. Also, four bits make a *nybble*. Some computers deal in 16 bits or even 32 bits. These larger bits can contain more information, build larger memories, and therefore handle larger and more complex tasks.

PROGRAMMING

The CPU's fixed features constitute its instruction set. This is its repertoire of routines which perform at the request of, perhaps, several dozen externally generated commands. These commands, in binary-number form, constitute a *machine language*. This rudimentary language has a precise, inflexible vocabulary, specific for each chip model. As with all languages, however that vocabulary can recombine in endless ways, giving the CPU's fixed routines great flexibility. Its overall performance depends ultimately on an external *program*, a step-by-step set of instructions.

A program is like a printed worksheet, such as, for example, an income-tax form. According to your circumstances and your income sources, the IRS requires you to process variable financial data through a minutely detailed procedure based on federal law, its instruction set. Here and there, the main form/program might direct you to a supplementary form, to detour, in effect, to a subroutine. Everything gets back on track, however, and, if your obedience is total and true, leads to an inevitable resolution: an offering to appease the IRS.

The whole form-filling procedure is so inflexible that a machine could do it, you might think. Of course, you would be right. Such a machine is a computer. Just feed it the raw data, and it will fill out any forms it can read. The machine-readable form is a program, a detailed procedure to sequence the computer's set of reflexes, its instructional set. Within the instruction set's limitations, programs can control any definable procedure, from tax compliance to flight trajectories.

Programming proper comes from other devices. For example, some chips which serve the CPU contain fixed circuits that function as programs. They may be special-purpose programs for controlling, say, a video display or a disk drive. Their fixed circuits are termed *firmware*. Firmware may also contain an applications program, such as a game, and so may come in replaceable cartridge form. Task-oriented programs, however, usually come on an erasable medium, such as tape or disc, and is called *software*. You can buy preprogrammed software, or you can prepare your own software with some higher-level language, which is intermediate between machine and human languages. All programs exist in some accessible form of storage, known as *memory*.

Memory

Digital memory comes down to registering sequences—machine-readable sequences of opposites—in some stable medium. Almost any medium will do, provided that the sequences endure.

Computers have special chips for *volatile* memory, referred to as *random access memory* (RAM). This is the readily accessible temporary storage area for programs. These same chips hold the newly created files before you save them to disk. Memory size is measured in bytes. As already noted, a byte can represent a binary number; a byte can also represent a character, a punctuation symbol, a space, and so on. There are about 200 words per 1000 bytes, or 1K, of memory. A 64K (64,000 bytes) memory can hold over 10,000 words. Actually, memory size increases in powers of two. So 1K is really 1,024, or two to the 10th power, and 64K is really 65,536, or two to the 16th power.

The information stored in RAM is lost as soon as power is removed. To permanently save a program or file, an alternate form of memory is necessary. Magnetic tape, diskettes, and disks are currently the most popular forms of permanent storage. In each of these media, the sequence of opposites are created with magnetic fields that are more or less permanent in nature. Commercially available programs are also generally available on one of these three media.

Somewhere between RAM memory and permanent memory are specialized circuits called *read-only memory* (ROM). The circuitry in ROM is similar to the circuitry in RAM, but the program or file originally placed in ROM is permanent. It can be accessed by a computer, but it cannot be changed.

Data Transmission

Data bytes move from place to place in either of two general ways: serial or parallel transfers. In *serial transfer*, the bits move one at a time over a single wire; so the sender must transmit a disassembled byte one bit at a time, and the receiver must subsequently reassemble it. In *parallel transfer*, the whole byte moves at once over eight lines, one line for each bit. Other things being equal, parallel transmission is quicker than serial, but serial transmission, having simpler and less bulky lines, can cover greater distances.

Even short distances within the computer involve a major transmission-line problem: The many possible linkups cannot have individual lines. They must share common lines, called *bus lines.* The computer selectively isolates the bus lines for transmission between pairs of switching points, whether serial or parallel.

Clearly, outside the computer, both serial and parallel transmission require a compatibility between sender and receiver. The devices must keep in step and distinguish between consecutive bytes with some method of synchronization. They may require additional bits to separate the bytes or check for accuracy.

Signal Codes. Numbers make excellent symbols for digital communications. Used thus they extend a long evolution of signaling. The earliest signaling over distance used sight or sound (smoke, beacons, drums, flags, semaphore, etc.), often in relays. Each message involved translating from language to code, and from code back to the original language, itself a code. Semaphore had the highest information potential, because it used separate symbols for all the letters of the alphabet, and was therefore capable of combinations without limit.

Later, the electric telegraph also coded individual letters, using the Morse code in patterns of short and long electrical pulses—dots and dashes. The receiving end had some audible or visual device which reacted to hand-keyed transmissions. Initially, human operators did the coding and decoding through their hands, ears/eyes, and trained reflexes. Machines eventually took over, however, evolving into ticker-tape machines, teletypewriters, and teleprinters. The machines eventually outgrew the Morse code and adopted, long before digital computers, a more efficient and larger capacity five-digit binary code. Eventually, more comprehensive codes became necessary as more varied equipment, uses, and users proliferated.

Information Interchange Codes. Digital computers manage most data transmission with fewer than 128 distinct characters. These include uppercase and lowercase letters, numerals, punctuation symbols, and special control commands. So, a code based on seven binary bits could cover most typical digital exchanges. The most extensively used code like this is the American Code for Information Interchange, *ASCII* (pronounced "as-kee"). The following are some ASCII character codes:

27	(0011011)	ESC	(Escape)
36	(0100100)	$	(Dollar Sign)
53	(0110011)	5	(Five)
63	(0111111)	?	(Question Mark)
78	(1001100)	N	
110	(1101110)	n	
125	(1111001)	}	(Right Brace)
127	(1111111)	DEL	(Delete)

Some of the items apply only to synchronous communications and teletypewriter-like devices. A 6-bit, 64-item version exists which omits lowercase letters, and a less accepted 256-item version also exists. Note that, although the normal code uses only seven bits, the computer moves them around as part of an 8-bit unit.

Another character code, used extensively in IBM mainframe equipment, is the Extended Binary Coded Decimal Interchange Code, or *EBCDIC*. This code covers the same commands as ASCII but represents them with different values. This incompatibility may prevent a personal computer from directly accessing the data base of a corporate computer.

Differently coded text files and equipment using different codes can also exchange information. They do this through automatic *translation tables*, provided by software or firmware.

Coding starts at the computer keyboard. Each key actuates a momentary-contact switch. So pressing keys, singly or in combination, triggers minicircuits to release binary-coded sequences of pulses.

SUMMARY

This brief survey has touched on most of the basics upon which the later chapters build. Its aim has been to bridge the gap between the assumed grasp of the typical personal-computer user and the perhaps scary gimmickry of useful peripherals. The user cannot always draw upon or afford expert advice. So, ahead of time, the prospective buyer must do some homework. We hope that these preliminaries encourage the adventurous to probe deeper.

3

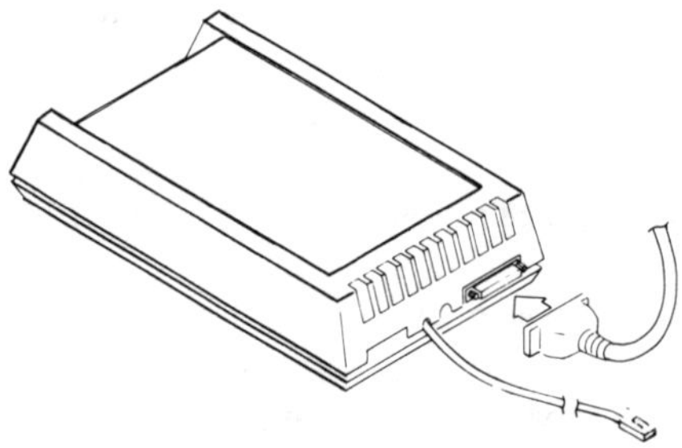

The Modem

Even the earliest computers needed to exchange data. As long as they were near each other, connecting them with cables was relatively easy. As their use and versatility grew, however, connecting widely separated computers presented a major problem; stringing cables between every branch office and corporate center simply was not cost-effective. One solution that seemed promising entailed using the existing telephone system. The American telephone system, perhaps the world's best, already connected nearly every potential computer site. Why not just connect the computers to the telephone system and transfer data through it?

This sounded like a perfect solution. It had some major defects however. The phone system had developed strickly for voice communication; it responds best to the oscillating type waveforms of sound, and only to the most effective portion of the voice-frequency range. It cannot accept directly the square-wave pulses of the computer's digital output. The phone system presents too good an opportunity to miss however. All that is really needed to use it is a means to convert direct current (dc) pulses into oscillation pulses within the phone system's range of audio frequencies and, subsequently, to convert those oscillation pulses back into dc pulses. Converting from dc to oscillations is called *modulation*. Converting oscillations to dc is called *demodulation*. The device that both modulates and demodulates is called a *modem* (Fig. 3-1).

SERIAL TRANSMISSION

Another major obstacle to transmitting data over phone lines concerns the way digital data is coded. As described in Chapter 1, computers represent each character by a data byte, a pattern of seven or eight logical 1s and 0s called bits. These bits exist in the computer as on or off electrical charges or current flows. Generally, the computer transfers the bits simultaneously over a set of conductors called a bus. Each phone, however, connects to the phone system with only two wires, forming an electrical loop. So, without resorting to

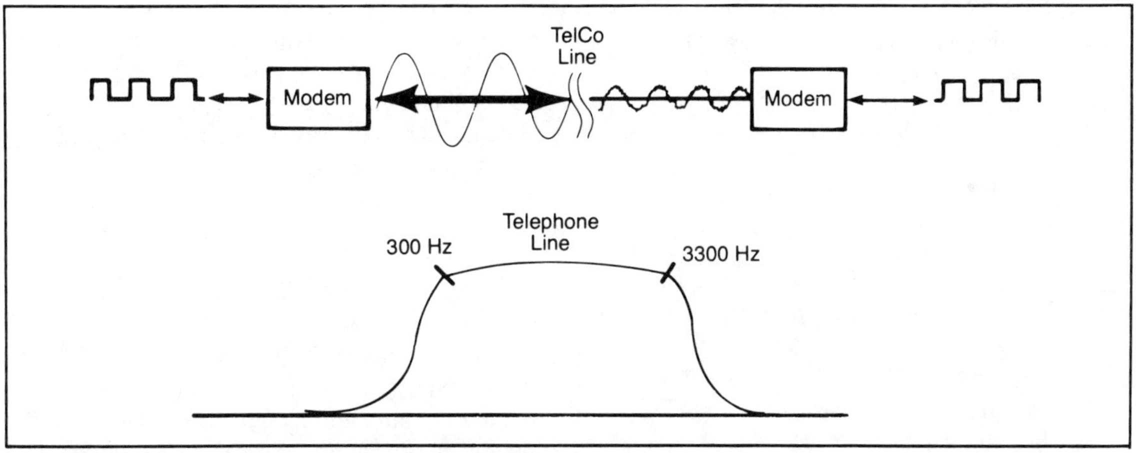

Fig. 3-1. Modem is a contraction of the words "Modulator-Demodulator" and is an electronic device used to convert digital signals to analog form for transmission over the telephone network. Courtesy Racal-Vadic.

additional lines, the bits must enter the existing phone loop serially—one bit after another in a single stream.

Consequently, before modulation, the bytes must be disassembled into individual bits. The device which does this is called a *communications adaptor*, or *serial card*. Some computers already have a communications adaptor; others require the insertion of a separately purchased device. The reverse process—the reassembly into the original byte pattern—occurs in the receiving communications adaptor. In between, the modems modulate and demodulate the bits by either of two processes: *frequency shifting* or *phase shifting*.

Frequency Shifting

The slower and less costly modems use frequency-shift keying (FSK). This process converts the digital two-level flat pulses into corresponding two-tone soundless beeps. It alternates between a pair of frequencies. A *mark* (logical 1) pulse becomes the upper frequency, and a *space* (logical 0) pulse becomes the lower frequency (Fig. 3-2). At the receiving end, the beeps revert to flat pulses.

The simplicity of frequency-shift keying keeps down modem prices. It works well enough up to 600-baud (600 bits per second) transmissions. Faster switching of frequencies meet certain

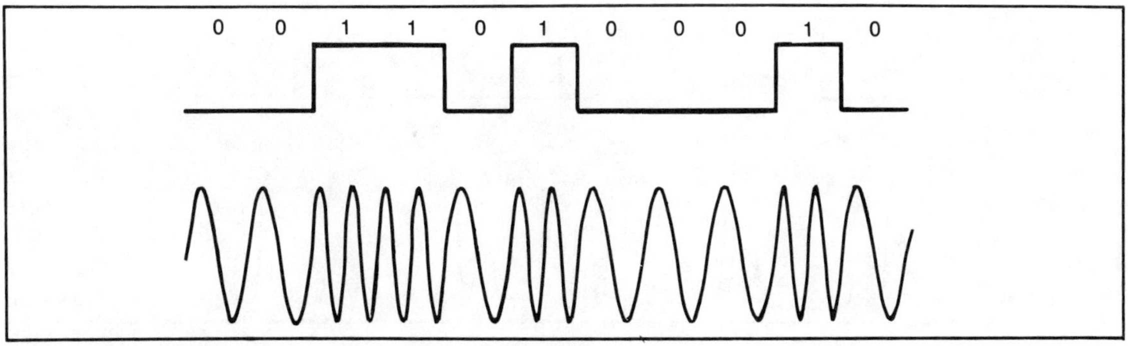

Fig. 3-2. Frequency modulation is used by the Bell 103/113 series for full-duplex, two-wire transmission at speeds of 0 to 300 bps. The Bell 202 series also uses frequency-modulation transmission. The transmitter shifts from one frequency to another as the data changes from a 1 to a 0, or a 0 to a 1. Courtesy Racal-Vadic.

natural limits. The upper FSK limit with full-duplex transmissions (see section on Duplex in this chapter) is about 1200 baud.

Phase Shifting

Faster modems modulate and demodulate with the more complex phase-shifting keying (PSK). A phase shift amounts to adding or deleting part of an oscillation cycle of a single frequency. It is rather like a marcher doing a little hop to change step. The phase-change, however, occurs every other bit by an amount which depends on the bit transition—00, 01, 10, or 11 (Fig. 3-3). At the receiving modem, the beeping reverts to dc pulses.

The sophistication of phase-shifting circuitry raises modem prices significantly, but such modems can operate at 1200 baud or greater. Some modems have both frequency shifting and phase shifting, using frequency shifting only for the slower speeds. Of course, only compatible modems can interact.

Duplex

Both methods of transmitting waveform pulses, instead of flat-level pulses, have a surprising advantage: within limits, each phone line can carry multiple frequencies simultaneously. This ability permits tuning to a particular channel, like radio and TV, while ignoring other channels. So, a modem can receive on one frequency, or pair of frequencies, while transmitting on another frequency, or pair of frequencies. This simultaneous two-way capability, built into many modems, is called *full-duplex*. Note that telephone conversations, in the sense of being simultaneously two-way, are also full-duplex. By contrast, *half-duplex* refers to one-way-at-a-time transmissions, like CB radio. Some half-duplex modems can send only, some can receive only, and some can do both. The half-duplex modems that can do both cannot begin sending until they receive an end-of-transmission signal.

When two modems attempt full-duplex exchange, it must be certain that they are each transmitting on different frequencies. That is, the modems must assume contrary modes, often called *originate* and *answer* modes. These terms are slightly misleading; they have nothing to do with which modem initiates the communication. Depending on the modem standard, the originate mode transmits on one frequency, and the answer mode transmits on another frequency. Where a choice exists, the modems must agree on modes before transmission. One or both terminals may, however, have permanently set modes.

Synchronization

Serial transmission must avoid confusing consecutive bytes. If, for some reason, a particular byte should lose or gain a bit, all of the subsequent data bits would reassemble into bytes incorrectly. Therefore, the sender must indicate the proper

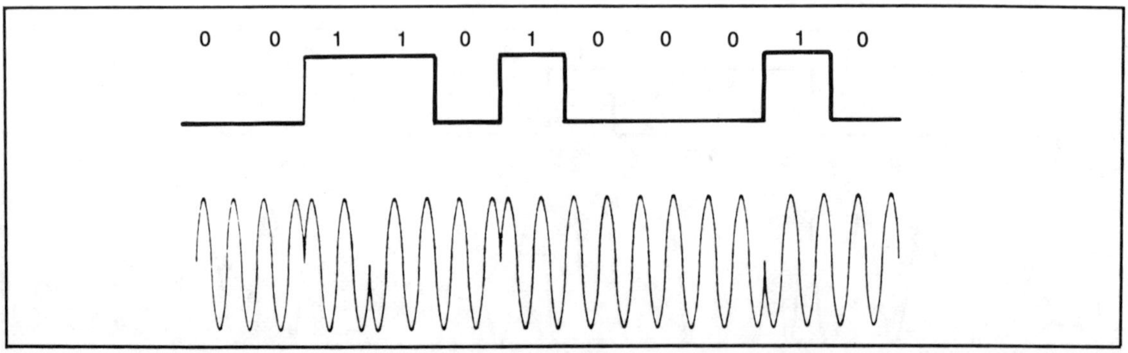

Fig. 3-3. As the data rate increases, more sophisticated modulation techniques are needed. The most common is phase-shift keying or dibit phase-shift keying. In this example, the phase of the carrier is changed once for every two bits of information, hence the term "dibit" phase-shift keying. The receiving modem must be able to detect the phase change of the carrier. Courtesy Racal-Vadic.

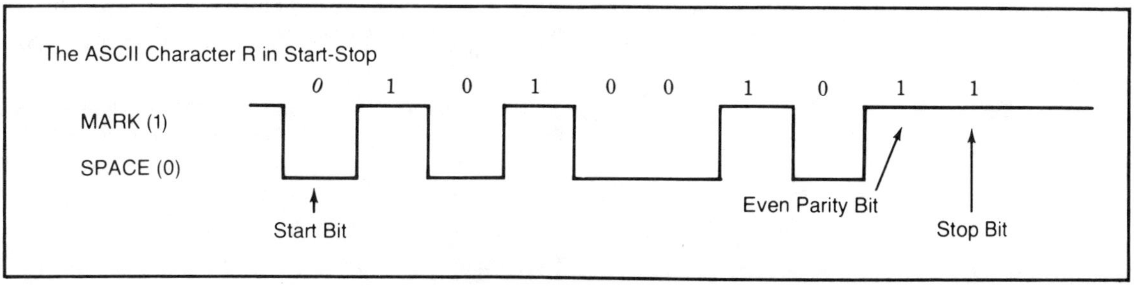

Fig. 3-4. Asynchronous operation is shown. Courtesy Racal-Vadic.

separation, confining a bit to its own byte. For this purpose, modems might use either synchronous or asynchronous transmission.

Asynchronous transmission frames each byte of data with additional start and stop bits (Fig. 3-4). These bits allow the receiving device to recognize the beginning and ending of each byte. Moreover, the asynchronous mode can transmit at irregular rates.

Synchronous transmission inserts special characters before a long series, or *block*, of data bytes (Fig. 3-5). These characters indicate the block beginning and may also indicate the block length. Usually, a separate clock signal between the devices paces the transfer. Therefore, the receiving device knows exactly when each new bit should arrive. A separate set of characters marks the ending of the block and may provide additional information concerning the accuracy of the transmission. Although more complicated, asynchronous transfer is potentially faster than synchronous transfer.

INTELLIGENT MODEMS

Some modems go beyond simply converting digital pulses from the computer into frequencies that can travel over the phone lines. By including a microprocessor and some memory, they can provide some very sophisticated features which are handy for specific applications. Usually, these additional features interact with the communications software running on the computer; that is, the modems hardware cooperates with the computer's software. Following are some of the more useful features on these intelligent modems.

Auto Dial

This feature allows a computer to reach another system without dialing the number manually. You enter the number through the computer keyboard and either the modem or the computer records the number in memory. Upon request, the signals associated with the number enter the phone lines and effectively dial the number. Phones use one of two types of signals for dialing: pulse or tone. All rotary dialing is pulse dialing; the number of pulses depends on the digits dialed. Some push-button phones are pulse dialing also; they have pulses stored in memory. Tone dialing uses buttons to release memory-stored tones—two tones for each digit. Usually, an intelligent modem can either pulse or tone dial, but the type of dialing must be the same that your regular phone uses. The telephone

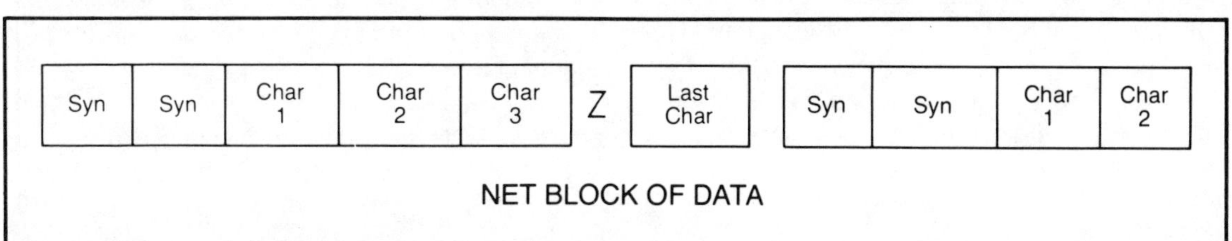

Fig. 3-5. Synchronous operation is shown. Courtesy Racal-Vadic.

company can change the dialing type for a standard charge.

Auto Answer

Some modems can automatically answer an incoming call. This feature allows data transmission to occur with an unattended system. Auto answer forms an integral part of electronic mail systems, where a user can call up an unattended computer and leave a message. It also allows commercial database systems and bulletin boards to release information automatically.

Self Test

When a data transmission fails to occur as expected, locating the problem becomes a rather complicated process. The failure could originate in any of several places: the computer itself, the software running on the computer, the interface to the modem, the phone lines, the modem at the other end, the other interface, the other computer, or the software running on the other computer. To verify the operating conditions, some modems provide a self test. To also help locate the source of any problems, some provide either an analog-loopback test, a digital loopback test, or both. The combined results of both kinds of test can more effectively isolate the defective part.

Analog Loopback Test. An analog loopback effectively connects a modem's output to its input. Data leaves from a terminal and passes through the modem. Instead of entering the phone lines, however, it feeds back into the originating modem. This returned data reflects any problem with the computer, the computer's software, the interface, or the modem itself.

Digital Loopback Test. Unlike an analog loopback, a digital loopback requires the assistance of the operator at the other end. The distant modem loops back your data before it reaches the other computer. Errors in this returned data indicate a problem with your system or modem or the distant modem.

Internal Speaker

Some modems include internal speakers which enable the user to monitor the progress of calls. For example, with auto dial, the speaker sounds the ringing or busy signal at the other end. The speaker also enables the user to hear either the carrier frequency of the other computer or a voice response. Although many intelligent modems often display status information on the screen, such as RINGING or CONNECTED, a speaker alerts the user for the unexpected.

LED Displays

Some external modems provide a series of *light-emitting diodes*, or LEDs, to monitor the status of the modem. Typically, a *power-on* LED indicates that the modem is receiving proper voltage. Often, a *carrier-detect* LED indicates that the modem has detected the distant carrier frequency, verifying contact. Additional LEDs may indicate the sending or receiving of data. They often help to diagnose problems with data communications.

MODEM ARRANGEMENTS

The number of variations and variety of modems probably rivals the number of available computers. Most of the features touted by modem manufacturers interest those with specific applications in mind. An understanding of the various features will help you to determine if they benefit your application.

Depending on design, modems mount either inside or outside the computer. An *internal modem* mounts directly into the computer's internal expansion board. It has the advantage of minimizing cable clutter and not taking up additional desk space. Being always designed for a specific computer, an internal modem will work with that computer with virtual certainty, most likely taking full advantage of that computer's potential. It will work only with other computers of nearly identical design. Internal modems often include necessary software.

An *external modem* has its own housing and remains outside the computer. It connects through a cable to the communications adaptor and should work with any computer with a proper RS-232C interface (see Chapter 4). It does not take up any of the computer's expansion slots. Because it is readily removable, and not computer specific, the external modem will work with a new or upgraded system. Most external modems require the purchase of additional software.

COMMUNICATIONS SOFTWARE

Perhaps the most important feature included with some modems is communications software. For a modem to interact with a computer, some form of software must be running on the computer. Software so importantly affects data communications and modems, we have dedicated the second part of this book to the subject. Nearly every feature of a modem must interact with the communications software, requiring a great deal of compatibility between modem and program. Some modems, particularly internal modems, include communications software, which ensures compatibility and that the software will fully use the modem's and the computer's potentials.

4

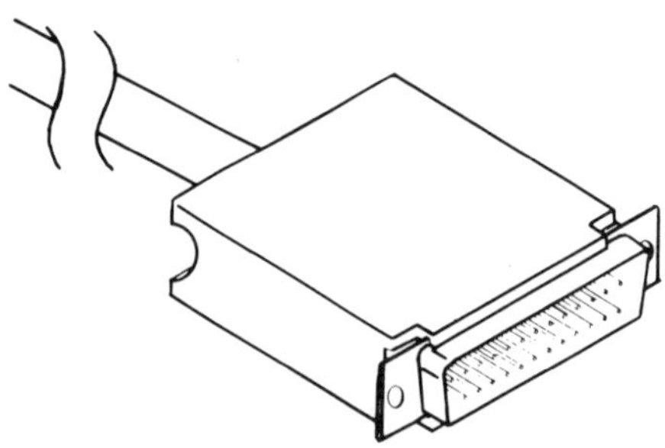

Modem Standards and RS-232C

For each modem that modulates, another compatible modem must demodulate, so modems need some sort of design standard. The earliest work in modems occurred at Bell Laboratories, which developed a variety of modems. For the most part, Bell modems set the standard for most United States manufacturers. These standards prescribe exactly how modems should modulate and demodulate and how to interact properly. Another widely used modem standard originated with the Racal-Vadic Company. Various other countries have established modem standards by a treaty called the CCITT, the French initials for International Consultative Committee for Telephone and Telegraph. The CCITT standards are similar to, but not interchangeable with, Bell standards.

BELL STANDARDS

Following is a list of Bell Telephone modem standards, used by most American manufacturers. Note that Hz is the abbreviation for hertz.

Bell 103. Bell 103 is the most common modem standard for full-duplex transmissions up to 300 baud. It uses the frequency-shift process described in Chapter 3, with 1070 Hz and 1270 Hz for originate mode, and 2025 Hz and 2225 Hz for answer mode (Fig. 4-1). A single integrated chip contains much of the circuitry, keeping the overall modem price under $100.

Bell 113. Bell 113 resembles Bell 103 in design, but it lacks full-duplex capabilities. Some models can send only; some can receive only, and some can either send or receive, but not at the same time (Table 4-1). Bell 113 uses the same lower and upper channels as Bell 103 and has the same 300-baud limit.

Bell 202. Bell 202 is a 1200-baud half-duplex modem. Some models can send only; some can receive only, and some either send or receive, but only one way at a time. It uses frequency-shift keying, but, unlike the slower frequency-shifting modems, it uses the same two-frequency channel for either sending or recieving: 1200 Hz and 2200 Hz.

Bell 212A. Bell 212A has become one of the most common standards for 1200-baud full-duplex transmission (see Fig. 4-2 and Table 4-2). It also

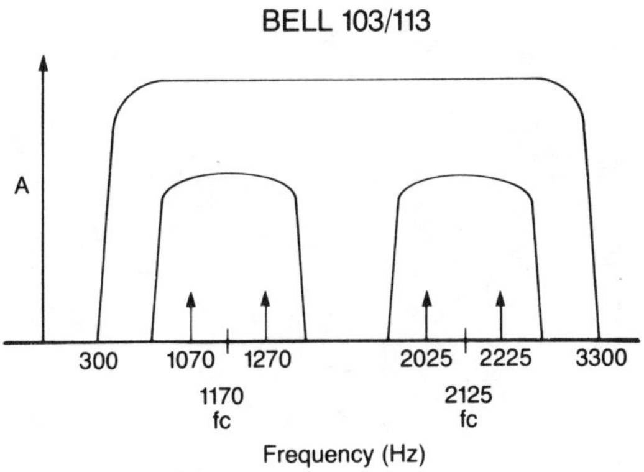

SPECIFICATIONS

Data:

Serial, binary, asynchronous, full duplex

Data Transfer Rate:

0 to 300 bps

Modulation:

Frequency shift-keyed (FSK) FM

Frequency Assignment:

	ORIGINATING END	ANSWERING END
Transmit	1070 Hz SPACE 1270 Hz MARK	2025 Hz SPACE 2225 Hz MARK
Receive	2025 Hz SPACE 2225 Hz MARK	1070 Hz SPACE 1270 Hz MARK

Transmit Level:

0 to −12 dBm

Receive Level:

0 to −50 dBm simultaneous with adjacent channel transmitter at as much as 0 dBm.

Fig. 4-1. Specifications and channel assignments for the full-duplex, 300 bps, asynchronous Bell 103/113 modem are shown in this illustration. Courtesy Racal-Vadic.

specifies the capability of running at 300 baud in a Bell 103 format. At 1200 baud it uses the phase-shifting process, described in Chapter 3, with 1200 Hz for originate and 2400 Hz for answer. At 300 baud it uses frequency shifting and the frequencies of Bell 103.

VADIC 3400

Vadic 3400 was the first 1200 baud full-duplex modem used on ordinary phone lines. (See Fig. 4-3 and Table 4-3.) It too can transmit from 0 to 300 baud full-duplex, but it uses phase shifting at all speeds. In the originate mode it transmits at 2250

Table 4-1. Bell 103/113 Family, 0 to 330 bps.

DIAL-UP	LEASED LINE
103A2, A3 OR J Originate/answer Single channel, self-contained	**103F** Originate/answer Single channel, self-contained
103E Data Station — Computer Site (Obsolete) Originate/answer 40 channels max	
113A/D Terminal Data Set Originate only Single channel, self-contained	
113B/C Data Station — Computer Site Answer only 20-channel increments, 240 max	

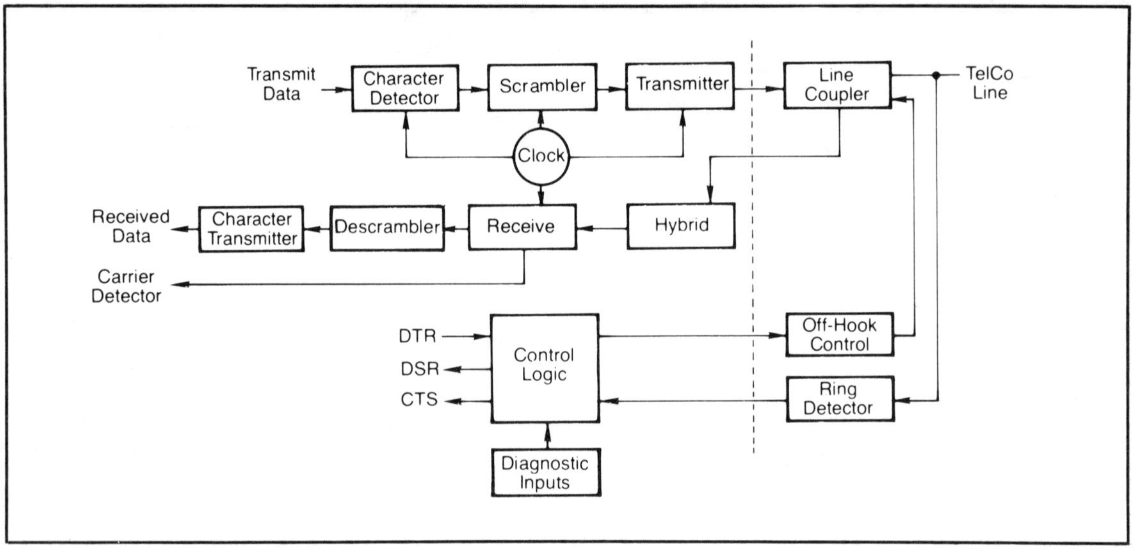

Fig. 4-2. The 212A from the Bell system is a full-duplex, 1200 bps modem for use on dial-up lines. Courtesy Racal-Vadic.

Hz; in the answer mode it transmits at 1150 Hz. The Vadic 3400 standard has received less attention than the Bell 212A standard, but most of the national commercial data bases still provide compatibility with both.

CCITT STANDARDS

As noted earlier, the CCITT standards were established to govern modems used outside the United States. They include the V.21, which is similar to the Bell 103, for 0 to 200 baud; the V.22, for 1200 baud; and the V.21, which is similar to Bell 202, for 600 baud and 1200 baud. Although the international modems resemble the Bell modems, they are not interchangeable; a Bell 103 modem cannot communicate with a CCITT V.21 modem.

Table 4-2. Bell 212A Specifications.

SPECIFICATIONS

Data:

Serial, binary, asynchronous or synchronous, full duplex

Data Rate:

1200 bps asynchronous or synchronous 0 to 300 bps asynchronous via FSK 103

Mode:

Full duplex on dial-up lines

Modulation:

1200 bps — dibit phase-shift keyed (DPSK)
300 bps — frequency-shift keyed (FSK)

Channel Assignment (1200 bps):

	ORIGINATING END	ANSWERING END
Transmit	Fc=1200 Hz	Fc=2400 Hz
Receive	Fc=2400 Hz	Fc=1200 Hz

Phase Shift:

DIBIT	212A
00	90°
01	0°
10	180°
11	270°

Channel Assignment (0 to 300 bps):

	ORIGINATING END	ANSWERING END
Transmit	1070 Hz SPACE 1270 Hz MARK	2025 Hz SPACE 2225 Hz MARK
Receive	2025 Hz SPACE 2225 Hz MARK	1070 Hz SPACE 1270 Hz MARK

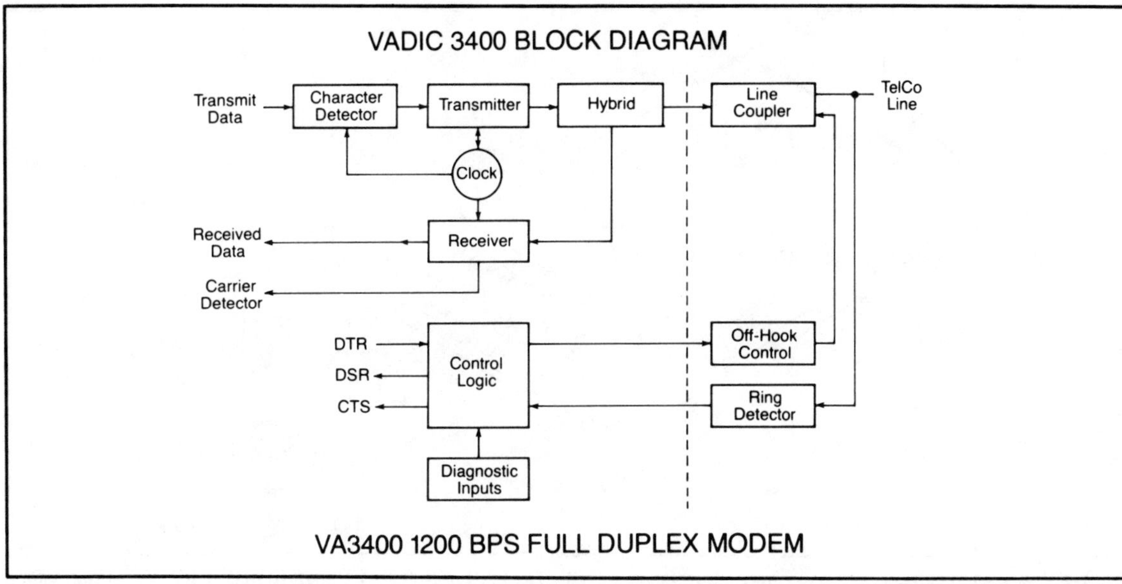

Fig. 4-3. The VA3400 is the world's first full-duplex, 1200 bps modem for use on dial-up or two-wire leased lines. Courtesy Racal-Vadic.

THE RS-232C INTERFACE

As you recall, computers represent data as patterns of voltages. We refer to the patterns as bytes, and to the voltages as bits. Before a byte of data can reach a modem, a communications adaptor must disassemble the byte and pass it on serially, one bit at a time. The modem must know how to receive them; so some sort of agreement must exist on how to transfer those individual bits. The devices must agree on voltage and current levels; they must agree on when to transmit data and when to receive, and both devices must recognize the first and last bit of a byte. Out of this need came the RS-232C interfacing standard (Fig. 4-4).

Defined by the Electronics Industry Association, the RS-232C has become the most common

Table 4-3. VA3400 Specifications.

SPECIFICATIONS

Data:

Serial, binary, asynchronous or synchronous, full duplex

Data Rate:

1200 bps asynchronous/synchronous or any speed 0 to 300 bps
600 bps asynchronous/synchronous or any speed 0 to 150 bps

Mode:

Full duplex on dial-up or 2-wire leased lines

Modulation:

Dibit phase-shift keyed (DPSK)

Transmit Level:

0 to −15 dBm

Receive Level:

−15 to −50 dBm (or 0 to −35 dBm) simultaneous with adjacent channel transmitter at up to 0 dBm

Channel Assignment:

	ORIGINATING END	ANSWERING END
Transmit	Fc=2250 Hz	Fc=1150 Hz
Receive	Fc=1150 Hz	Fc=2250 Hz

Phase Shift:

DIBIT	VA3400
00	90°
01	270°
11	180°
10	0°

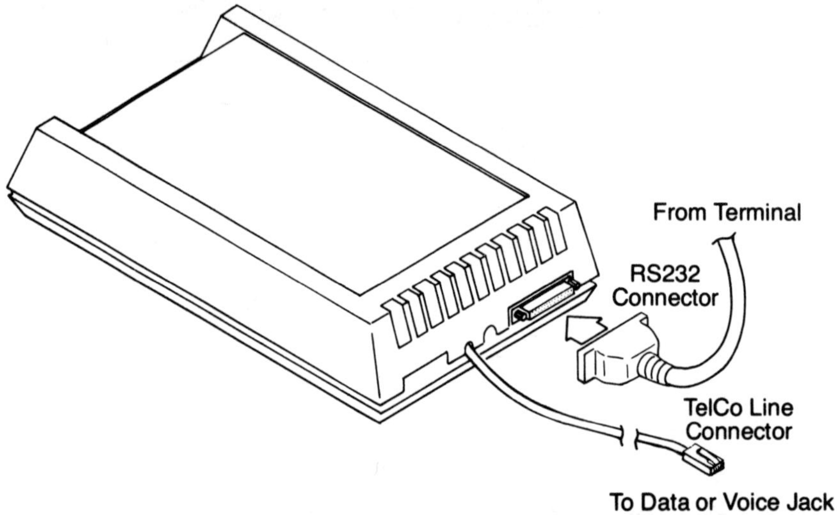

Fig. 4-4. A modem has a connection, commonly referred to as the RS-232C interface, for the terminal, and a connection for the telephone line. Courtesy Racal-Vadic.

standard for serial transmission with modems and printers. The standard outlines in specific terms the voltage levels for signals exchanged between devices. It describes 21 different lines that control the exchange of data, and it actually defines 13 possible configurations for using those lines. The various modem standards (e.g., Bell 103) designate how data will be transferred between two modems. The RS-232C standard designates how data is transferred between modem and computer.

The RS in RS-232C stands for *recommended standard.* In practice, however, few of the recommended configurations get used. Many devices use only three of the lines. Some devices ignore the definitions and use the lines for other purposes. Unfortunately, this happens to most "accepted" standards and severely complicates compatibility.

The standard limits the maximum length of the cable for connecting devices to 50 feet; however, many applications use cables up to 500 feet. Beyond 50 feet, cable capacitance significantly reduces the data transfer rate. Although the standard does not specify what type of physical connectors to use, the DB-25 connector serves the standard most often. Again, though, only three of the DB-25's 25 pins actually get used (Fig. 4-5).

Terminal and Communications Equipment

The standard defines four groups of signals: ground, data, control, and timing. Two lines carry data—one transmits data (TD), and one receives data (RD). The transmitting line of one device must connect to the receiving line of the other, and vice versa, so installers need to differentiate between the devices. For this, they usually use the terms *data terminal equipment* (DTE) and *data communications equipment* (DCE). A DTE device, such as a computer terminal, assigns pin 2 of the RS-232C cable as the TD line, and pin 3 as the RD line. A DCE device, such as a modem, does just the opposite. Ordinarily, therefore, a DTE device and a DCE device can connect directly. Occasionally, however, two DTE devices need to connect directly, such as when a computer terminal needs to connect to a host computer. A special cable, called a *null modem,* circumvents the problem by crossing the TD and RD lines internally.

Control Signals

The RS-232C standard defines nine separate control lines for the primary channel. (A secondary channel also provides data and control lines, but it is seldom used.) The voltage level for a control on

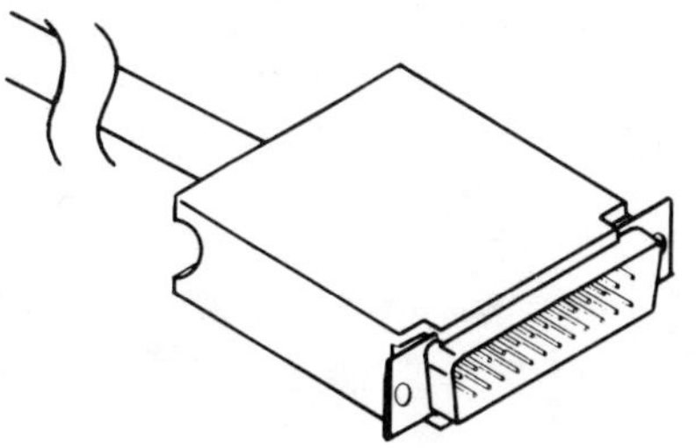

Pin	Function	Pin	Function	Pin	Function
1	Frame Ground	10	Negative dc Test Voltage	19	Sec. Request To Send
2	Transmitted Data	11	Unassigned	20	Data Terminal Ready
3	Received Data	12	Sec. Data Carrier Detect	21	Signal Quality Detect
4	Request To Send	13	Sec. Clear To Send	22	Ring Indicator
5	Clear To Send	14	Sec. Transmitted Data	23	Data Rate Select
6	Data Set Ready	15	Transmitter Clock	24	Ext. Transmitter Clock
7	Signal Ground	16	Sec. Received Data	25	Busy
8	Data Carrier Detect	17	Receiver Clock		
9	Positive dc Test Voltage	18	Receiver Dibit Clock		

Fig. 4-5. The terminal connection to the modem is defined by the Electronic Industries Association as specification RS-232C, which specifies the use of a 25-pin connector and the pin on which each signal is placed. Courtesy Racal-Vadic.

signal (other than data lines) must fall in the range of +5 to +15 volts. A control off signal must fall in the range of – 5 to – 15 volts. Again, seldom do all of the defined lines get used. Some of the more commonly used lines include *Request to Send* (RTS), *Clear to Send* (CTS), *Data Set Ready* (DSR), and *Data Terminal Ready* (DTR).

Data Terminal Ready. The DTR line lets a DTE (data terminal equipment) device indicate that it is ready. For example, a computer terminal uses this line to tell the modem that it is on.

Data Set Ready. The DSR line lets a DCE (data communications equipment) device indicate that it is ready. For example, a modem may activate this line when it is ready to transfer data. A communications program may examine this line before it attempts to transfer data to the modem.

Request to Send. The RTS line originates from the DTE device (e.g., a computer terminal) and indicates a desire to transmit data.

Clear to Send. A CTS signal comes from the DCE device (e.g., a modem) to acknowledge the RTS signal and to indicate readiness to receive data.

The following control lines get used to varying degrees, depending on the type and sophistication of equipment:

Carrier Detect (CD). As discussed earlier, modems convert data pulses into audio frequencies. A modem may contain circuitry that detects the frequencies from another modem, indicating that contact has been established. A CD line acknowledges this contact and sometimes illuminates a panel light or displays a message on a screen.

Ring Indicator (RI). Some modems contain circuitry that detects the ringing of a phone on an incoming call. A RI line indicates that the phone is ringing, and it sometimes initializes an auto-answer feature.

Signal Quality Detector (CG). A CG line becomes active when a device senses excessive noise that may interfere with the transfer of data.

Data Signal Rate Selectors (CH and CI). Some devices have settings for multiple rates of data transfer. The CH and CI lines sometimes serve to select between two rates.

Timing Signals

A fourth set of lines defined by the RS-232C standard enables devices to exchange timing signals without interfering with data transfer. Two *Transmit-Clock* (TC) lines and one *Receive-Clock* (RC) line function exclusively on synchronous transmissions, generally involving higher rates of transfer.

The Secondary Channel

The RS-232C standard defines a secondary channel with five additional lines for data transfer and control. This secondary channel has its own equivalent of transmit and receive lines, as well as RTS, CTS, and DCD lines. As noted, however, in practice, this channel is seldom used.

OTHER INTERFACE STANDARDS

In 1977, the Electronics Industry Association established a new standard for serial transmission. Actually, it included a group of three standards: RS-422, RS-423, and RS-449. The signals used resemble those of the RS-232C, but, unlike RS-232C, each signal has its own separate return line. This feature increases the number of wires necessary in the connecting cables, but it reduces interference, allowing transmission rates of 10 million bits per second and cable lengths beyond 4000 feet. The connectors, completely defined for the new standard, avoid some of the ambiguities in the RS-232C standard. The new standard, while used in some high-speed applications, has not received serious acceptance by manufacturers of modems for personal computers and microcomputers.

BEYOND THE STANDARDS

Although the RS-232C clearly defines hardware requirements, it leaves unanswered a number of questions involving the transfer of data. These are important questions, such as how fast to transfer data, how many bits will comprise a word, and how to recognize the separation between bytes of data. Sending and receiving devices must agree on these and other parameters, setting them prior to transmission through hardware and software controls.

Baud Rate. Some devices can transmit and receive data faster than others. The DTR, DSR, RTS, and CTS only facilitate data transfer; they do not provide a complete means of controlling the rate. Therefore, to avoid loss of data, both devices must agree in advance to a specific rate of transfer. Called *baud rate,* this value indicates the number of bits transferred per second.

Duplex. The RS-232C Interface, with separate lines for transmitting and receiving, allows simultaneous two-way or full-duplex signaling. This process occurs often between modems and terminals. Some applications, however, particularly with printers, transmit in half-duplex; or one direction at a time.

Start and Stop Bits. With serial transmission, the receiver needs to know where one byte of data begins and another ends. To alleviate this problem, the sender inserts start and stop bits between characters. Commonly, either one or two stop bits get used, but again, the transmitting and receiving devices must agree on the number prior to transmission.

Word Length. As discussed in Chapter 2, the ASCII code represents alphanumeric characters with seven-bit bytes, or words. Other codes, however, may use bytes with a different number of bits. Because serial transmission involves the disassembly and reassembly of bytes, agreement on byte length becomes vital.

Parity. Any type of electrical transmission can incur interference, or noise, which causes errors. For example, electromagnetic radiation from strong power sources within a piece of office equipment, a poor connection at an interface, or a faulty component within a device can pollute data. A common method of detecting this type of error uses a parity check. An even parity check counts the number of 1 bits in a byte, and, if necessary, adds another 1 to make the total even. If, on examination at the end of transmission, the byte no longer has an even number of 1s, the receiver requests that the byte be retransmitted. An odd parity check does the same for an odd total. Clearly, the devices must agree on either an even or odd parity check.

5

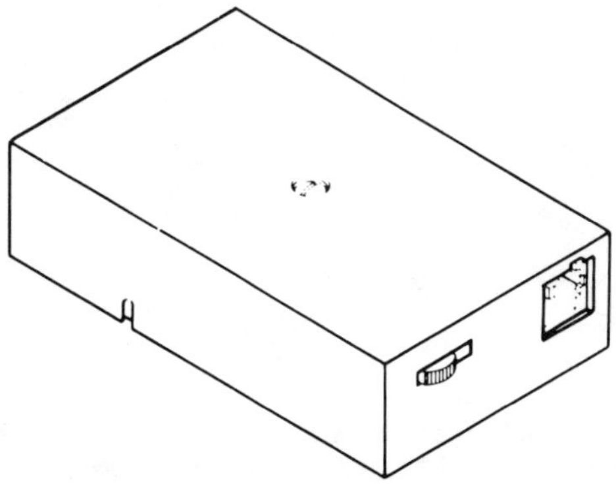

The Phone Connection

Problems when transferring data over the phone system become easier if you have some background on the system's structure. The present public telephone system was first designed in the early 1930s to send and receive voice signals only. Individual phone sites link to central offices, or exchanges, by two wires which circulate the current bearing the voice signals for both directions. Every call goes first to its local exchange. If the called phone shares the same exchange, then the exchange makes a direct linkup. If not, then the exchange reroutes the caller to the appropriate local exchange.

The Switched Network. Each local phone exchange constantly switches connections to establish contact between customers. For this reason, the public phone system is sometimes called the *switched network*. The network often reroutes calls through two or more local exchanges. For especially great distances, the data may travel through microwave links, with several repeater stations along the way to enhance the signal.

The constant switching and amplification often results in considerable interference, and even a call placed next door may have noticeable background interference. This noise can have a serious effect on the transmission; a receiving modem may confuse the interference for data. Transmissions at the higher rates of 1200 baud and above increasingly invite errors from interference. Beyond 4800 baud, data transmissions across public phone lines become extremely unreliable.

Private Lines. Many applications in business mandate the high-speed transmission of reliable data. Banking, for instance, which transfers values involving large sums of money, must keep data precise. So, instead of using switched networks, exacting applications may use private lines which avoid exchanges by connecting two sites directly. They may use the same type of cables as switched networks, or they may use superior cables, such as coaxial or fiber-optic cables. Leasing these lines is considerably expensive, but the lines provide quality transmissions at over 10,000 baud.

Packet-Switched Networks. In the mid seventies a number of firms introduced a hybrid of the switched networks and private lines. Sometimes

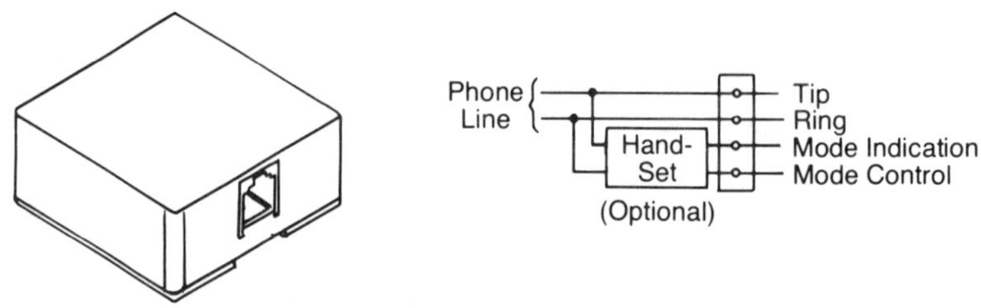

Fig. 5-1. Modems registered as "permissive" devices are most commonly connected to an RJ11C voice jack. This connection is used to connect a standard telephone to the telephone line. Permissive devices must limit the amplitude of the signal presented to the telephone line to a maximum of −9 dBm. Courtesy Racal-Vadic.

called *packet-switching networks,* these systems allow customers to connect, through the regular local exchanges, to nearby exchanges of the alternate network. The alternate route goes through leased lines, high-speed land lines, and satellite relays. A system of error detection and correction make this hybrid network superior to the completely public network. Two leaders in packet switching are Telenet and Tymnet. Many of the commercial data-base systems use these firms.

Modular Connections. At one time, new phone installations required a service call to physically wire the phone to the nearest outlet. More recently, phones are designed to accept a cable that simply plugs into a socket. These modular phones utilize a standard connector (RJ11C), which allows anyone to install a phone.

Modems utilizing RJ11C connections are registered as *permissive* devices; they contain at least one socket for connection to the wall outlet (Fig. 5-1). Some have a second socket for the regular phone which avoids disturbing the hookup for regular calls.

Another type of modular connection, RJ45S, is used on modems registered as programmable devices (Fig. 5-2). A programming resistor installed by the telephone company sets the transmit level of the modem to its optimum level, based on the distance

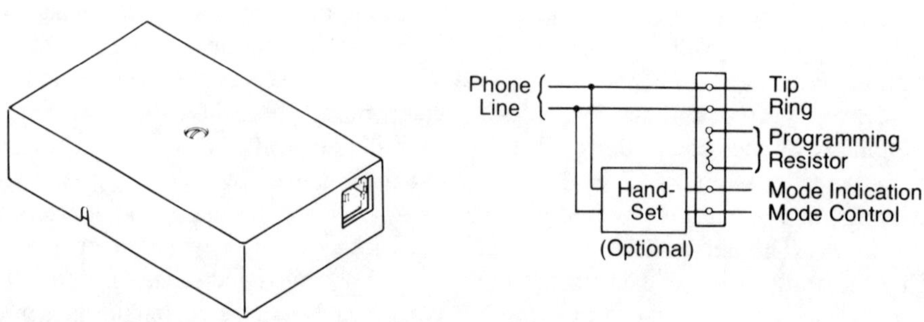

Fig. 5-2. Modems registered as "programmable" devices commonly connect to an RJ45S programmable data jack. Courtesy Racal-Vadic.

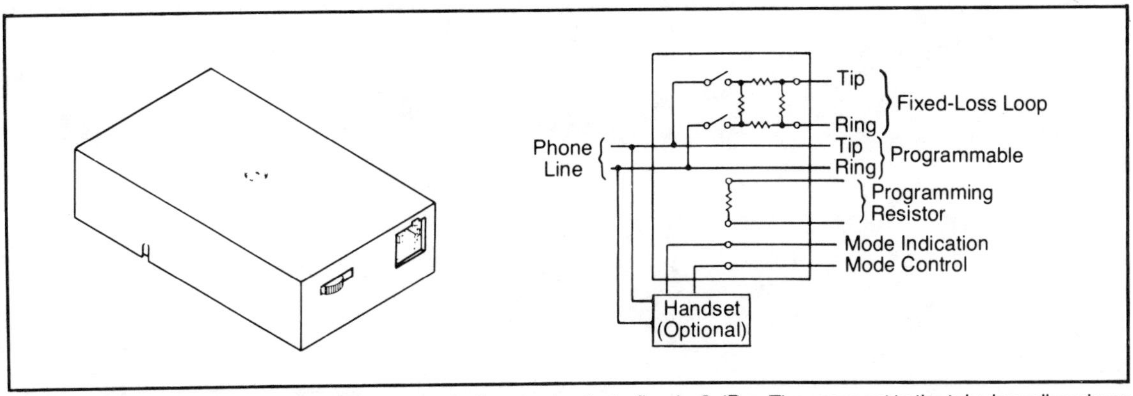

Fig. 5-3. Modems registered as "fixed-loss loop" devices transmit at a fixed −3 dBm. They connect to the telephone line via an RJ41S universal data jack. Courtesy Racal-Vadic.

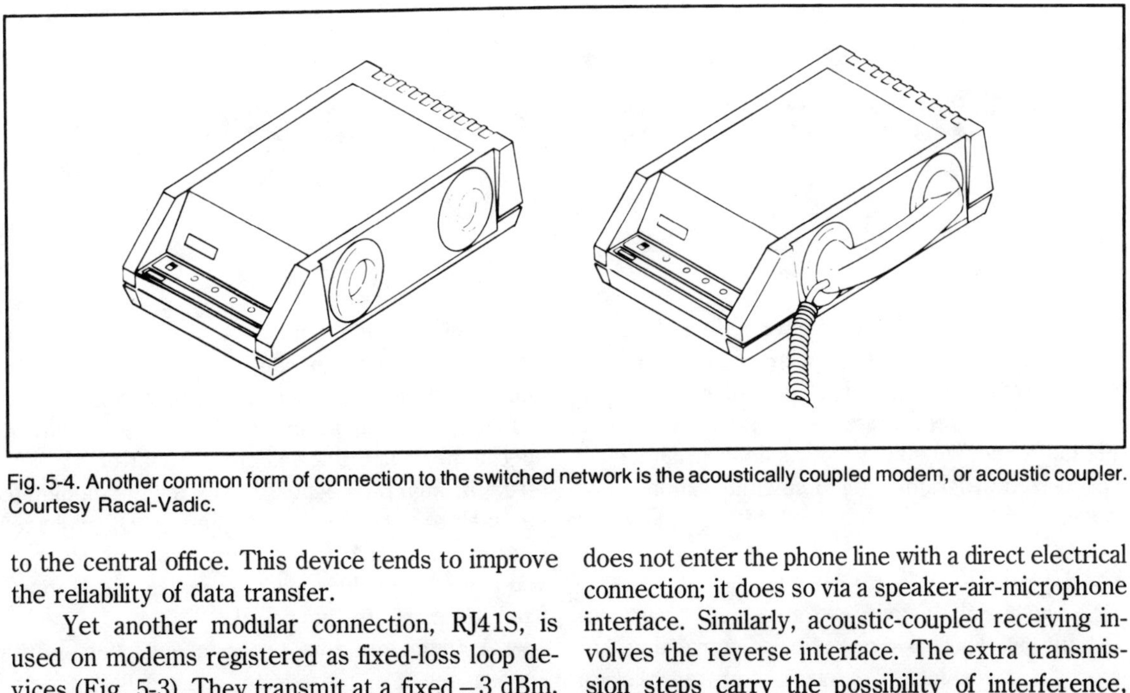

Fig. 5-4. Another common form of connection to the switched network is the acoustically coupled modem, or acoustic coupler. Courtesy Racal-Vadic.

to the central office. This device tends to improve the reliability of data transfer.

Yet another modular connection, RJ41S, is used on modems registered as fixed-loss loop devices (Fig. 5-3). They transmit at a fixed −3 dBm.

Acoustic Connections. Acoustic-coupled modems provide socketlike rubbery cups for inserting the handset of a telephone (Fig. 5-4). The mouthpiece of the handset inserts into one cup and the earpiece into the other cup. A data transmission does not enter the phone line with a direct electrical connection; it does so via a speaker-air-microphone interface. Similarly, acoustic-coupled receiving involves the reverse interface. The extra transmission steps carry the possibility of interference, particularly from extraneous sounds. These modems are handy for travel, because hotel rooms often avoid modular-type phones. As RJ11 modular connections have become more common, so acoustic couplings are used less.

6

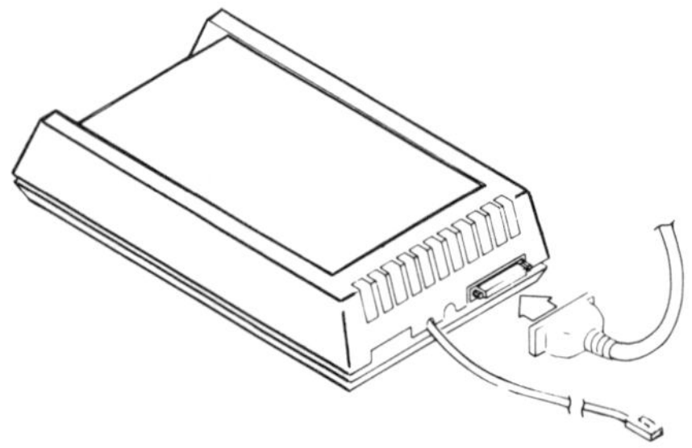

Communications Software

In spite of its potential, a computer without direction and control is useless. Similarly, an intelligent modem cannot transfer data without programmed instructions. Every interaction between a computer and a modem needs instructions provided by specialized programs, called communications software.

CONTROL OF THE MODEM

Much of the activity of the modem occurs automatically. The basic operation of converting digital information into frequencies for transmission across the phone lines requires no special attention from the user. Some of its functions, however, may require some intervention. Even the simplest of modems may require settings for either an originate mode or an answer mode.

As with most digital devices, the user exerts control through hardware or software settings. Hardware settings depend on a variety of switches, such as dual in package (DIP) switches, often used on circuitry boards.

Intelligent modems usually invoke their features through software control. When properly connected to a terminal, these modems recognize special characters, or groups of characters, as having special meanings. For example, the letter "A," entered through the keyboard, may tell the modem to set itself to answer mode; the letter "D" might tell the modem to shift to originate mode. In this way, you can control the modem's modes and features from the computer keyboard.

A communications program must contain the correct set of commands for a particular modem. For this reason, a communications program is often designed with a particular modem in mind. Many modems and communications programs use the set of commands originally developed by Hayes Microcomputer Products, Inc. The Hayes command set is becoming the unofficial standard within the industry.

Intelligent modems must function in two states: the *command* state, which recognizes commands from the keyboard, and an *on-line* state, which transfers data. Generally, a modem first enters the command state at startup. At that point, any character from the keyboard rates as a command and does not go over the phone lines. A special command places the modem into the on-line state. Here, characters from the keyboard become data, convert into frequencies, and enter the phone lines. In the on-line state, most command characters rank as just any other piece of data and transmit accordingly.

Interacting with communications software, control of a modem often works either through a menu selection or through English commands. In menu format, the screen displays a list of numbered options. Option selection results from simply keying in the appropriate number. In reality, the communications software issues the modem commands that activate that particular option, but the appearance is as if a single number had done it. Other programs allow English commands to effectively control a modem. For example, the word "DIAL" followed by a number appears to automatically dial that number. In fact, the software once again issues the modem commands necessary for the appropriate action.

Communications software varies in sophistication. The simplest software will transfer characters entered from the keyboard to the modem, and transfer characters received through the modem to the screen. This terminal software treats your computer as if it were nothing more than a keyboard and display unit. Exploiting the real data-communications potential of your computer calls for more demanding programming.

An important issue with communications software, more than with most types of programming, is compatibility. Communications software must, of course, work with the particular computer for which it is written, but it must also have some compatibility with other systems. For example, in transferring data to or from another system, it must agree on various parameters not defined in the RS-232C protocol, such as baud rate, duplex, and parity. It must also recognize the format and procedures used by other systems. For example, a computer system will often transmit extra characters that have special meanings to some terminals but which confuse personal computers. To avoid this confusion, some programs emulate a specific terminal by recognizing all the special characters. Other programs include methods of eliminating the special characters.

Each communications program operates in a variety of modes. Usually, a specific mode has a special function. For example, a *print mode* controls printing data, and a *transfer mode* controls transferring data. In different modes, each of the keys on the keyboard may take on new meanings. In the print mode, "P" starts the printer printing, but in the transfer mode, it may select a specific transfer protocol. Of course, different programs will use the keyboard differently.

Communications software must interact with the modem; it therefore operates in two fundamental modes: a command mode and an on-line mode. In the command mode, characters entered from the keyboard control the hardware or software. For example, the letter "D" may dial a number, and the letter "A" may set the modem to answer the phone. The command mode may also have submodes, such as the print mode and transfer mode, just described. In each case, however, the control operates through the keyboard. In the on-line mode, characters transfer from the keyboard through the modem to the host computer, but they do not exert any special control over any of the hardware of software.

PROGRAM DRIVE

Different communications programs have different features, but nearly all of them implement those features in either of two ways: by menu or by command. In a menu-driven program, the screen displays a list of numbered or lettered options. Keying in the character corresponding to a particular option implements that option. For example, if the list includes "P—PRINT," pressing the letter P causes data to print. Often a selection on a main menu triggers the display of a secondary menu. For example, if the main menu includes "C—CONFIGURATION," pressing the letter C will result in a second list of options applying to configuration. Occasionally, a menu selection produces a prompt for additional information. For example, pressing P for PRINT may display "ENTER FILE NAME." The user responds with a file name and a carriage return.

In command-driven programs, menus are conspicuously absent. You select options by entering appropriate commands from the keyboard. For example, entering through the keyboard "DIAL 566-1234," followed by a carriage return, may instruct the modem to dial the number. The command "BAUD 1200," followed by a carriage return, may set the modem to a speed of 1200 bits per second. The various commands are often displayed through a help screen. This screen usually provides a brief description and the command format for each command. The software manual describes the command's detailed effect.

Both menu-driven and command-driven modes have strengths and weaknesses. Menu-driven programs are easier to use. The operator can see the options at any time, and first-time users can readily use a menu with only minimal background in computers and data communications. Advanced users, however, may find viewing a series of menus irksome. Since most users of communications programs repeatedly select the same applications, they would prefer to command that application directly. Command-driven programs allow this, but they require a thorough understanding of the available commands. The computer novice may find the command structure and format confusing; however, the more advanced user exploits the command-driven program's versatility. For example, the user can tailor a program for a specific application, such as saving a list of commands for instant recall. Aspects of both menu-driven and command-driven controls combine in some of today's communications software.

MAIN FEATURES

In all its variety, communications software has essentially four major categories of features: configuring, dialing, answering, and transferring data. Intelligent modems perform many of the dialing and answering routines, but the software still exercises overall control.

Configuring

As mentioned in Chapters 4 and 5, modems transfer data according to the RS-232C interface standard. This standard defines many of the signals and circuits of data communications, but leaves unanswered a number of important questions, such as baud rate, word length, start and stop bits, word length, parity, and duplex. Before transmitting data, the transmitting and receiving devices must agree on values for these parameters. The values are usually set under software control through keyboard entries. These entries actually reset the hardware, just as if you pushed a button or moved a switch. Different computer systems have different parameter settings. Communications software lets the user change the setting conveniently and at will. Some programs allow you to save the settings on disk and to invoke those desired settings automatically.

As mentioned earlier, some computers are built to interact with a specific terminal. That terminal has special characters which affect the screen's data display, but which may confuse a different type of terminal or computer. Some, but not all, communications programs have methods of alleviating that problem. They may allow you to filter off unwanted characters or to change them to harmless ones. Some programs will automatically configure your computer to behave like a specific type of terminal, so it will recognize and use the special characters. This process of *terminal emulation* may be important when accessing an office computer. Most commercial data-base systems, however, make a big effort to be compatible with all types of computers.

Configuring may extend to connected printers, benefitting the printing of data received from another computer. It also allows the software to adjust the data signals to suit the type of printer connected.

Dialing

Some intelligent modems have the ability to dial a number automatically. This feature is usually implemented by transmitting to the modem a special dialing command followed by the number to be dialed. Many communications programs provide a mode for commanding the modem directly. Other programs further simplify the process by allowing the user to record numbers and names into a directory, allowing the operator to enter a single-character code from the directory to automatically execute the dialing of the appropriate number.

Some modems can recognize a busy signal and redial the number repeatedly. Still others, on recognizing a busy signal, can dial a second number. Again, these activities must be programmed. Communications software attempts to simplify the programming. For example, some menu-driven software will provide redialing as an option, and some command-driven programs provide redialing commands. Some of the dialing parameters, such as how long to let the number ring before giving up, or how many times to redial, may be set during the configuration of the software. Again, a program may provide a menu option or prompt you for a command.

Most commercial data bases require preliminary information from a computer attempting to access information. They charge for the information provided and, for billing purposes, require an identification number. The system prompts the caller to enter the number. Then the system asks for a secret password, known only to the caller, which prevents others from avoiding billing by guessing at an identification number. This preliminary logging on, or *logon,* is an inevitable part of accessing most systems.

Some software allows you to program keys to transmit identification numbers or passwords automatically; pressing a single key will completely answer a question. Some communications programs store logon information in special files which will automatically transmit an entire logon sequence after making a connection. This is a little trickier than it sounds; before answering, the software must wait for the questions. Some programs enable you to delay a response for a given amount of time. Others will actually match a word in a question to one stored in memory. For example, the program will wait until it sees NUMBER before sending the identification number.

Answering

Some communications software supports modems with auto-answer capabilities, allowing callers to an unattended system to view directories, upload and download files, and leave and collect messages. Usually associated with electronic mail or bulletin board services, a caller gains control to that system as if sitting at its keyboard. Some of the more sophisticated programs can be configured to demand passwords before giving access to the system. This ensures that unauthorized callers cannot acquire or destroy important data.

Transferring Data

The primary function of communications software is to transfer data. Nearly all software of this type has a terminal mode, which treats your computer as if it were merely a keyboard and video display unit. The terminal mode transfers data to a receiving modem as you enter the data at your keyboard. The data displays on your screen as the other system receives it. The terminal mode allows the user to interact manually with the other system. Data reaches the receiving system only as fast as the operator can type.

You often want to save data received from another computer. One data-saving procedure *captures* data in a temporary storage area called a *buffer*. Use of a buffer generally includes the ability to toggle the capture mode on and off to store only portions of the data. Data received in a terminal mode escapes beyond recall as it scrolls off the screen, but data received in a capture mode can be reviewed, even after it has scrolled out of view, and saved permanently on disk. Often, a program will include the ability to edit a buffer, adding to or deleting from previously captured text.

A second data-saving procedure stores the data to disk directly by *downloading*, avoiding the potential loss of data because of a power failure. It limits the transfer rate to the writing speed of the disk. However, often, the terms capturing and downloading are used interchangeably. In fact, most communications software combine the two processes. Data is captured in a buffer periodically and automatically downloaded to disk. During the transfer from buffer to disk, the computer cannot accept new data; any data arriving during that time gets lost. To avoid this loss of data, the combined process of capturing and downloading usually relies on a flow-control protocol called *XON/XOFF*. Most mainframe computers and commercial data-base systems recognize this protocol in which a set of special characters tells a system to stop or start transmission. When a receiving buffer becomes filled with data, its computer sends a special character to stop the other computer from transmitting, empties the buffer to disk, then sends a second character to allow the other computer to resume transmitting.

Transmitting a data file from your system is called *uploading*. This process transfers files from disk to the modem directly, or through an intermediary buffer. Occasionally, a mainframe computer limits the rate at which it can receive a file. Data uploading from a connected computer may be lost if it is transferred at the agreed upon baud rate. Some communications programs allow a *throttling* of data to slow it down to an acceptable rate.

Most computer systems recognize two major types of files: binary and ASCII. In uploading or downloading an ASCII file, special characters indicate the beginning and ending of the file. A binary file contains machine-code instructions, including characters which may be confused for beginning or ending markers. Most, but not all, communications software is capable of transferring both types of files.

As mentioned in Chapter 4, parity checking is a method of detecting errors incurred during transmission. A parity check is not completely reliable, however; it can overlook multiple errors. Most communications programs, therefore, transfer files with more sophisticated error-detection techniques. Each transfer protocol is usually unique for a particular communications program. Clearly, both systems involved in a file exchange must recognize the same protocol, which often means using the same software. Gaining general acceptance, however, is the *XMODEM* protocol. Many communications programs include it in addition to their own protocols.

7

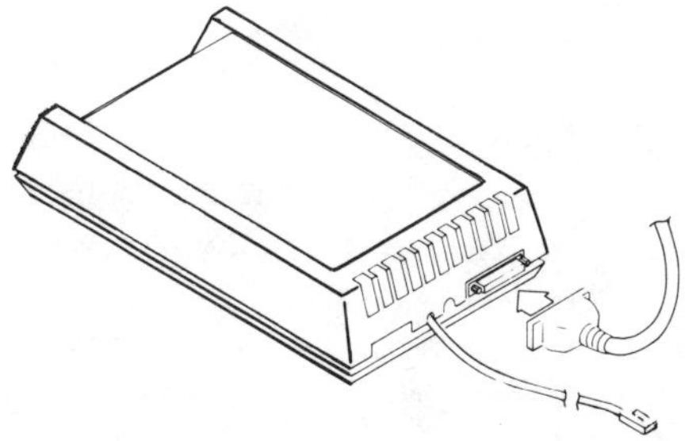

Selecting a Modem and Communications Software

This book aims to facilitate the selection of modems and communications software. Although it is not, perhaps, the last word, it is, at least, a starting point. This chapter presents some of the important issues you, as a consumer, should keep in mind as you select the modem and communications software you desire and need.

SELECTING A MODEM

When you attempt to select just one modem, the variety of modem types and available options can present a rather intimidating list. By a careful examination of your particular needs, you can, however, usually narrow that list down to a manageable few.

Applications. Perhaps the first thing to consider in selecting a modem is the application you have in mind. An application often requires specific modem features. If you plan to use your modem infrequently, or just for fun, you will probably get by with a 300-baud modem. If, however, you have business or commercial applications in mind, you will probably need a more expensive 1200-baud modem. Most of the national data bases discussed in Chapter 1 offer both 300- and 1200-baud transmissions, but the 1200 baud transmission often costs more. On the other hand, being nearly four times as fast, the more expensive rate may still yield considerable savings.

Applications involving electronic mail require modems with an auto-answer mode. Auto dial is worthwhile if you make frequent calls to certain numbers. If you plan to use the same phone for both data and voice communications, make sure that you can conveniently switch between data and voice modes. If your phone is the standard RJ11C modular type, definitely select a modem with that type of connection. If you travel constantly with a portable computer, you may, however, need an acoustic-coupled modem.

Computer Compatibility. Carefully read any documentation accompanying your particular computer concerning modems, expansion, RS-232C, and serial cards. Some computers provide empty slots within an internal expansion board which allows the addition of various options. Options generally include communications adaptors (serial cards), which will enable RS-232C devices to connect to the computer. Once inserted, a serial card can connect to an external modem. Often, internal modems designed for your particular computer may eliminate the need for a serial card. See Chapter 3 for a discussion of the pros and cons of internal and external modems.

Communications Software. When you buy hardware, the equally important software deserves top consideration. All modems need some form of communications software. At the very least, a program must transfer data from the computer keyboard to the modem. From there, the data must get to the screen for viewing. Some communications programs go far beyond these essentials. They facilitate the automatic dialing of numbers; they record and save received data; they simplify the setting of various protocol parameters, and, to verify that data has transferred correctly, they often manage a complicated error-checking scheme.

Many of the features you want may exist in the communications software, so the software must work properly with the modem you select. For that reason, you should purchase the modem, if at all possible, with the software in mind. Many modems, particularly internal modems, include communications software. This *bundled* package may be the safest way to select a modem and software. In many cases, it is the most economical. Many of the better communications programs are sold separately, however, and they provide some information about which modems are compatible with them. If in doubt, call the vendor to ensure compatibility.

Reviews. At the beginning of each review, we have listed the specifications important to anyone considering a modem, regardless of application. The format allows you to skim each review to determine a product's possible suitability. A brief description reveals specific features and limitations. Then a section goes into considerable detail on how various features work. In the last part of the book, we review software that comes with a modem.

We list the products in alphabetical order by manufacturer or distributor, including each vendor's address and phone number, a brief description of the firm, and the warranty and service agreement for their products. When you are selecting a product, service can be of considerable importance.

SELECTING COMMUNICATIONS SOFTWARE

The two most important issues in selecting communications software are compatibility with both the modem and the computer and application. On these matters, you should take particular care that the advertising means what you think it means.

Compatibility. Computer compatibility is vital. The software, of course, must run on your computer. Usually, the advertising clearly identifies computer compatibility, but occasionally, the specification says only that a software program has compatibility for a particular operating system, such as CP/M or MS/DOS, used by a number of computers. While in theory the program should run on any computer using the specified operating system, the manufacturers may deviate just enough with their computer to create problems for some programs. So before purchase, receive a guarantee from the vendor or retailer that a program will run on your system. The same situation develops with computers touted as IBM or Apple compatible. Make certain that you can return the communications program if it does not run on your computer.

Communications software must also be compatible with the modem. External modems work with any RS-232C compatible system, but this does not imply that they will work with any communications software. For instance, when a communications program attempts to simplify the process of dialing or answering, it must use commands recognized by the modem in use. The closest thing to a standard in modem commands is the Hayes command set. Many, but certainly not all, communications programs use these commands to ensure some compatibility. Often vendors of communications software will have complete lists of compatible modems. If any doubt remains about the compatibility of your modem and software, a long distance call to the vendor would be well worth the cost. The most substantial savings come from buying a modem and communications software marketed together in a single package. This costs less than buying them separately, and, without a doubt, it ensures compatibility.

A third level of compatibility may be of importance: compatibility with a remote system. Most commercial data-base information services ensure compatibility with a maximum number of computers, modems, and communications software. To access information, however, some corporate computers may require specific terminals. These terminals recognize the special screen-formatting characters that the mainframe sends. If you wish to access information with a computer other than the specific terminal, the communications software will have to emulate that terminal by recognizing the special characters. Only a few programs have this feature.

Data transfers between computers, particularly at above 300 bits per second, may need checking for errors. The technique for error detection is often unique for a particular communications program. Therefore, to ensure identical protocols, the signal convention for ordering a retransmission, two computers may need to use the same software. Numerous communications programs use an error-detection protocol called XMODEM. This protocol allows two different programs to detect for errors while transferring files.

Applications. A second consideration in selecting communications software is the application. The application needs determine the program features. Some program features may truly benefit the application, while others unnecessarily complicate and raise the cost of the software.

Useful for almost any application is the ability to download information, which allows the user to save information received from a host system. The information can then be read or printed at a later, more convenient time.

Another useful feature is dialing from a directory, which enables the user to avoid fumbling through a phone list and manually dialing a number. You just key in a single character corresponding to a directory listing, and the computer and modem automatically dial the number. This feature streamlines accessing a variety of frequently used services.

Automatic directory dialing, as well as a few variations, can greatly simplify accessing commercial data-base services. Accessing services such as The SOURCE, the Dow Jones News /Retrieval Service, and CompuServe, which provide current information on stocks, news, products, and events, is becoming a common application of personal computers. Each service, however, has its own protocol for logging on, for which you may often need to check the documentation. Providing the same essential logging-on information for each access becomes tedious when done frequently.

Some software, in addition to automatic dialing, allows automatic logging on to information services. This feature involves configuring the software with the logging-on protocols. Then, a single-character command can trigger the whole logging-on procedure. More sophisticated software goes further—it allows you to create programs to automatically access and save information. For example, with a single command you could automatically dial the Dow Jones News/Retrieval Service, request the current price of a particular stock, print the price on your printer, and disconnect from the data base. If your computer has a built-in clock, you could even arrange for the whole retrieval sequence to occur while you sleep or are absent.

Another unattended application, electronic mail, uses auto answering and a few special software features. With the appropriate modem, some auto-answering programs will allow a caller to access information in your computer. The caller can also leave a message for you to read later. With more sophistication, your software can require the caller to provide a password; it denies access to incorrect passwords. Software can also help the caller, providing either a menu or a set of commands.

A more common application involves user interaction with a distant system directly, in real time. This application requires a terminal mode, provided by nearly all communications software. In the terminal mode, data entered at the keyboard and displayed on the screen transmits directly to the receiving system. Similarly, the screen directly displays data sent from the other system. The terminal mode is useful for accessing bulletin boards and for communicating directly with another operator.

Reviews. In the section on features, we have briefly mentioned the various options and modes of operation. This section gives the user who is familiar with communications software a convenient way of determining suitability. For others, a review of some of the earlier chapters will refresh understanding about the various options and modes. The discussion section provides a more detailed account of the program's features. We often emphasize modes or options unique to the program.

If possible, the communications and modem should be selected together. You want to avoid paying for program features which the modem does not support. Equally wasteful would be modem features which the software does not use. So the most cost-effective combination will likely take some back and forth comparisons, keeping an eye on compatibility. The compatible modems and computers listed in the reviews come directly from the software documentation. The combined purchase of modem and software can represent a sizable investment. We hope that the review information and the manner of presentation clarify and simplify the selection process.

8

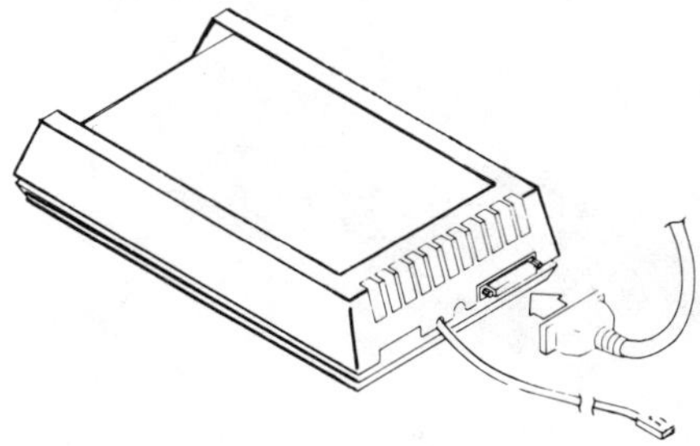

Modem Reviews

Reviews of modems are listed alphabetically by company. Prices are as of 1984.

ANCHOR AUTOMATION, INC.

6913 Valjean Ave.
Van Nuys, CA 91406
(213) 997-6493

Anchor manufactures a number of modems and cables for their modems. The company warrants equipment manufactured and sold by it to be free from defect in material and workmanship and to meet applicable specifications under normal use and service for a period of 2 years.

Models reviewed and suggested retail prices:

Volksmodem	$ 79.95
Signalman Mark I	$ 99.00
Signalman Mark III	$139.00
Signalman Mark X	$169.00
Signalman Mark XII	$399.00

Volksmodem

BAUD: 300
COMPUTER: Any RS-232C-compatible system
TYPE: Bell 103
SOFTWARE: Not included
AUTO ANSWER: No
AUTO DIAL: No
PHONE: RJ11 (modular) connection

Features. The Volksmodem provides two modular sockets for connection to a compatible phone and to an appropriate wall outlet. Two switches on the modem toggles between full-duplex mode and half-duplex mode and between phone mode, for normal conversation, and data mode, for data transfer (Fig. 8-1).

The additional purchase of a special interface cable is necessary. Available cables include an RS-232C cable, with a DB25 connector, for connection

Fig. 8-1. The Volksmodem 300-baud modem. Courtesy Anchor Automation, Inc.

to systems compatible with RS-232C. Other cables include direct connection to the user port of the Commodore 64.

The Volksmodem is powered by a 9-volt battery (not included). A single cable for connecting the phone to the modem is included. Communications software, necessary for using the modem, must be purchased separately.

Discussion. Installation of the Volksmodem is quick and convenient. When the modem is set to data mode, a short, high-pitch tone indicates that the battery is working. In the talk mode, drain on the battery is reduced, and you can use your phone for normal conversation.

The Volksmodem senses the carrier of the modem at the other end and automatically selects either the answer or originate mode. The duplex setting on the modem must agree with software settings. Under most situations, the setting is left in full-duplex. All other parameters must be set through communications software, now provided with this modem. A tone from the modem alerts the operator to a proper connection with a host computer.

Signalman Mark I

BAUD: 300
COMPUTER: Any RS-232C-compatible system
TYPE: Bell 103
SOFTWARE: Not included
AUTO ANSWER: No
AUTO DIAL: No
PHONE: RJ11 (modular) connection

Features. The Mark I has two modular-phone (RJ11C) sockets—one for connection to an appropriate phone outlet, and one for connection to a modular phone. The DB25 connector from the modem permits connection to any RS-232C interface with a like connector. It is presumed that a connected computer will have appropriate software for implementing the RS-232C standard.

A single panel switch selects between voice mode, where the phone is used for normal conversation, and data mode, where it is used for data transfer.

The Mark I requires a 9-volt battery (not included), for internal power. A single cable for connecting the phone to the modem is included. Communications software necessary for using the modem must be purchased separately.

Discussion. Installing the Mark I is quick and convenient. A modular cable, included with the modem, connects the base of the phone to the modem. Another cable connects the modem to an appropriate phone outlet. The modem is battery powered. When the battery is low, the modem beeps at one-second intervals. An optional 9-volt dc adapter allows a connection to a 115-volt ac wall socket.

This modem automatically selects either an answer or originate mode, depending on the modem at the other end of the line. Calls are dialed manually in the voice mode. When a carrier frequency is heard at the other end, you must change the panel switch to data mode. A tone from the modem indicates that it is locked on to the remote carrier and ready for data transfer.

The documentation provides little more than assistance in installation and troubleshooting.

Signalman Mark III

BAUD: 300
COMPUTER: Texas Instruments 99/4A
TYPE: Bell 103
SOFTWARE: Not included
AUTO ANSWER: No
AUTO DIAL: No
PHONE: RJ11 (modular) connection

Features. The Mark III has two modular-phone (RJ11C) sockets—one for connection to an appropriate phone outlet, and one for connection to a modular phone. The DB25 connector from the modem permits connection to any RS-232C interface with a like connector. It is presumed that a connected computer will have appropriate software for implementing the RS-232C standard.

A single panel switch selects between voice mode, where the phone is used for normal conversation, and data mode, where it is used for data transfer.

The Mark III requires a 9-volt battery (not included), for internal power. A single cable for connecting the phone to the modem is included. Communications software, necessary for using the modem, must be purchased separately.

Discussion. Installing the Mark II is quick and convenient. A modular cable, included with the modem, connects the base of the phone to the modem. Another cable connects the modem to an appropriate phone outlet. The modem is battery powered. When the battery is low, the modem beeps at one-second intervals. An optional 9-volt dc adapter allow connection to a 115-volt ac wall socket.

This modem automatically selects either an answer or originate mode, depending on the modem at the other end of the line. Calls are dialed manually in the voice mode. When a carrier frequency is heard at the other end, you must change the panel switch to data mode. A tone from the modem indicates that it is locked on to the remote carrier and ready for data transfer.

The documentation provides little more than assistance in installation and troubleshooting.

Signalman Mark X

BAUD: 300
COMPUTER: Any RS-232C-compatible system
TYPE: Bell 103/113
SOFTWARE: Not included
AUTO ANSWER: Yes
AUTO DIAL: Yes (pulse- or tone-dial)
PHONE: RJ11 (modular) connection

Features. The Mark X is a 0- to 300-baud modem compatible with Bell 103 and Bell 113 (Fig. 8-2). It is capable of full-duplex or half-duplex operation. It incorporates a microprocessor and associated firmware to provide auto-dial and auto-answer. The modem can be used with most communications programs that use Hayes modem commands.

The Mark X has two modular-phone (RJ11C) sockets—one for connection to an appropriate phone outlet, and one for connection to a modular phone. The DB25 connector from the modem permits connection to any RS-232C interface with a like connector. It is presumed that a connected computer will have appropriate software for implementing the RS-232C standard.

Dialing commands enable either pulse or

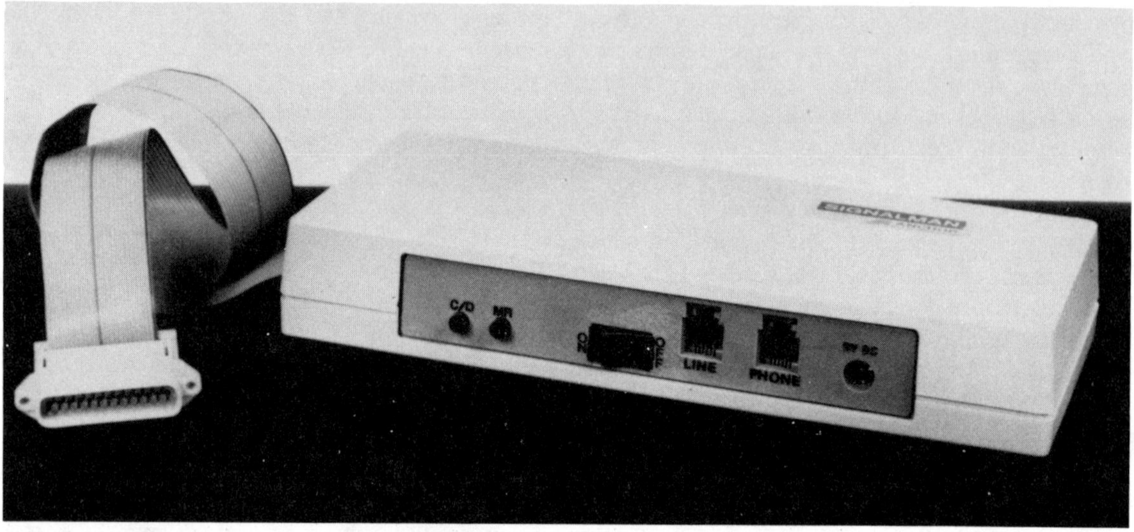

Fig. 8-2. The Signalman Mark X modem. Courtesy Anchor Automation, Inc.

touch-tone dialing, automatic redialing, and delays between numbers (when a second dial tone exists).

Auto-answer commands can set the modem to answer the phone after a predetermined number of rings. Result codes, such as CONNECT, RING, BUSY, DIALING, DIAL TONE, and NO CARRIER, all of which monitor the progress of calls, can be displayed on your terminal's video-display unit.

Discussion. Installing the Mark X is quick and convenient. The base of the phone is connected to the modem with a modular cable, included with the modem. The modem is then connected to an appropriate phone outlet. The modem includes a 9-volt dc adapter for connection to a 115-volt ac wall socket. A single power switch applies power to the modem.

The modem assumes one of two states: the command mode or an on-line mode. The modem assumes the command state when first powered up. One- or two-character commands are entered from the keyboard. The commands are used to set RS-232C parameters, such as baud rate and duplex, and also to implement the various dialing instructions.

The on-line mode consists of two submodes: answer and originate. The originate mode places calls either manually or automatically. The answer mode accepts incoming calls either manually or automatically.

The modem will disconnect automatically if it is unable to obtain a carrier frequency from a remote modem within 18 seconds after the initial connection. The modem will also disconnect automatically if, after making the connection, the carrier frequency is lost for more than ½ second. The data link can be disconnected manually, either by sending the appropriate command from the terminal, or by unplugging the connection with the wall outlet.

Programmable registers within the modem allow you to set the ASCII value to represent the carriage return, the line feed command, the backspace character, and the escape code. An LED on the front panel indicates that power has been applied, and the modem is ready to operate. A second LED indicates that the modem has detected a valid carrier from a distant modem.

Signalman Mark XII

BAUD: 1200/300
COMPUTER: Any RS-232C-compatible system
TYPE: Bell 212A/103/113
SOFTWARE: Not included
AUTO ANSWER: Yes
AUTO DIAL: Yes (pulse- or tone-dial
PHONE: RJ11 (modular) connection

Features. The Mark XII is compatible with Bell 212A and Bell 103 and has operating speeds of 1200 baud and 0 through 300 baud (Fig. 8-3). It incorporates a microprocessor and associated firmware to provide auto-dial and auto-answer. The modem works with most communications programs which use Hayes modem commands.

The Mark XII has two modular-phone (RJ11C) sockets—one for connection to an appropriate phone outlet, and one for connection to a modular phone. The DB25 connector from the modem permits connection to any RS-232C interface with a like connector. It is presumed that a connected computer will have appropriate software for implementing the RS-232C standard.

Dialing commands enable either pulse or touch-tone dialing, automatic redialing, and delays between numbers (when a second dial tone exists). Auto-answer commands can set the modem to answer the phone after a predetermined number of rings.

Result codes, such as CONNECT, RING, BUSY, DIALING, DIAL TONE, and NO CARRIER, all of which monitor the progress of calls can be displayed on your terminal's video-display unit.

Discussion. Installing the Mark XII is quick and convenient. A modular cable, included with the modem, connects the base of the phone to the modem. Another cable connects the modem to the appropriate phone outlet. The modem includes a 9-volt dc adapter for connection to a 115-volt ac wall socket. A single power switch applies power to the modem.

The modem features two states: the command mode and an on-line mode. The modem assumes the command state when it is first powered up. One- or two-character commands are entered from the keyboard. The commands are used to set RS-232C parameters, such as baud rate and duplex, and also to implement the various dialing instructions.

The on-line mode consists of two submodes: answer and originate. The originate mode places calls either manually or automatically. The answer mode accepts incoming calls either manually or automatically.

The modem will disconnect automatically if it is unable to obtain a carrier frequency from a remote modem within 18 seconds after the initial connection. The modem will also disconnect automatically if, after making the connection, the carrier fre-

Fig. 8-3. The Signalman Mark XII modem. Courtesy Anchor Automation, Inc.

quency is lost for more than ½ second. The data link can be disconnected manually, either by sending the appropriate command from the terminal or by unplugging the connection from the wall outlet.

Programmable registers within the modem allow you to set the ASCII value to represent the carriage return, the line feed command, the backspace character, and the escape code.

LEDs on the from panel indicate various conditions:

- The modem has power and is ready to operate.
- It has detected a valid carrier from a distant modem.
- Data is being transmitted or received.
- It is operating at 1200 baud.

APPLE COMPUTER, INC.

20525 Mariani Ave.
Cupertino, CA 95014
(408) 996-1010

Apple, one of the first companies to successfully market personal computers, also manufactures modems and serial cards for their computers.

Models reviewed and suggested retail prices:

Apple Modem 300	$225.00
Apple Modem 1200	$495.00

Apple Modem 300

BAUD: 300
COMPUTER: Any RS-232C-compatible system
TYPE: Bell 103
SOFTWARE: Included
AUTO ANSWER: Yes
AUTO DIAL: Yes (pulse- or tone-dial)
PHONE: RJ11 (modular) connection

Features. This external 300-baud modem includes a special cable for connecting to a standard DB-25 connector. The back of the modem includes two RJ11C sockets for connecting to a wall outlet and a phone, a series of rocker switches for configuring the modem, a power switch, and a power outlet for connecting a power adapter included with the modem.

The modem can dial numbers from keyboard entries through touch-tone or pulse-dial environments. Various commands tell the computer to pause between dialing instructions, redial the last number, display messages from the modem, and echo commands. The small internal speaker allows you to monitor the call and listen for busy signals and wrong numbers. In the auto-answer mode, the modem will answer an incoming call after a programmed number of rings.

Discussion. The Apple Modem operates in two functional states: the command state and the on-line state. In the local command state, ASCII characters each have a special meaning. For example, the character "D" followed by a number will instruct the modem to dial that number. The character "," will instruct the modem to pause for two seconds to wait for a second dial tone before dialing the outside number. In the command state, entries from the keyboard travel only as far as the modem; they are not transmitted over the phone lines to another computer.

The Apple Modem goes into the on-line state after answering a call or dialing a call and connecting with a distant modem. In the on-line state, characters are transferred normally to a receiving system through the phone lines. The character "d" is transferred as the letter itself, and it no longer causes the modem to dial a number. Entering a special "escape" character will put the modem back into the command mode. Any character can be assigned as an escape character through a special command sequence described in the documentation.

A rocker switch behind the modem provides a data terminal override if your terminal does not supply the data terminal ready signal. A second switch will let you set the carrier-detect line upon detection of a carrier frequency.

The Apple Term program included with the modem only works with the Apple Super Serial Card. The program has four display screens: main menu, terminal mode, configuration mode, and update mode. This menu-driven program allows you to create a phone directory of numbers that will automatically call the specified number using the RS-232C parameters necessary for that system. Data can be keyed in from the keyboard and transmitted to the receiving system in the terminal mode.

Apple Modem 1200

BAUD: 1200/300
COMPUTER: Any RS-232C-compatible system
TYPE: Bell 212/103
SOFTWARE: Included
AUTO ANSWER: Yes
AUTO DIAL: Yes (pulse- or tone-dial)
PHONE: RJ11 (modular) connection

Features. This external 1200-baud modem includes a special cable for connecting to a standard DB-25 connector. The modem is Bell 212-compatible at a baud rate of 1200, and Bell 103-compatible at baud rates of 0 to 300 baud. The back of the modem includes two RJ11C sockets for connecting to a wall outlet and a phone, a series of rocker switches for configuring the modem, a power switch, and a power outlet for connecting a power adapter included with the modem.

The modem can dial numbers from keyboard entries through touch-tone or pulse-dial environments. Various commands tell the computer to pause between dialing instructions, redial the last number, display messages from the modem, and echo commands. The small internal speaker allows you to monitor the call, listening for busy signals and wrong numbers. In an auto-answer mode the modem will answer an incoming call after a programmed number of rings.

Discussion. The Apple Modem 1200 operates in two functional states: the command state and the on-line state. In the local command state, ASCII characters have special meanings. For example, the character "D" followed by a number will instruct the modem to dial that number. The character "," will instruct the modem to pause for two seconds to wait for a second dial tone before dialing the outside number. In the command state, entries from the keyboard travel only as far as the modem; they are not transmitted over the phone lines to another computer.

The modem goes into the on-line state after answering a call or dialing a call and connecting with a distant modem. In the on-line state, characters are transferred normally to a receiving system through the phone lines. The character "d" is transferred as the letter itself; it no longer causes the modem to dial a number. Entering a special "escape" character will put the modem back into that command mode. Any character can be assigned as an escape character through a special command sequence described in the documentation.

A rocker switch behind the modem provides a data terminal override if your terminal does not supply the data terminal ready signal. A second switch will let you set the carrier detect line upon detection of a carrier frequency. A third switch will compensate for nonstandard phone lines caused by some PBX or CBX systems which do not meet Bell specifications.

The Apple Term program included with the modem only works with the Apple Super Serial Card. The program has four display screens: main menu, terminal mode, configuration mode, and update mode. This menu-driven program allows you to create a phone directory of numbers that will automatically call the specified number using the RS-232C parameters necessary for that system. Data can be keyed in from the keyboard and transmitted to the receiving system in the terminal mode.

BUSINESS COMPUTER CORPORATION (BIZCOMP)

532 Weddell Dr.
Sunnyvale, CA 94089
(408) 733-7800

Bizcomp holds the patent on command-driven modems, used by most modems with auto-dial features (Fig. 8-4). They are currently licensing the technology to other manufacturers, including Hayes Microcomputer Products, Inc.

Bizcomps products are warranted against de-

Fig. 8-4. Bizcomp's command-driven modems. Courtesy Business Computer Corporation.

fects in materials and assembly for 2 years from the date of delivery. Software is warranted for 90 days.

Models reviewed and suggested retail prices:

1022 Intelligent Modem	$279.00
1012 Intelligent Modem	$549.00
PC:IntelliModem	$499.00

1022 Intelligent Modem

BAUD: 300
COMPUTER: Any RS-232C-compatible system
TYPE: Bell 103/113
SOFTWARE: Not included
AUTO ANSWER: Yes
AUTO DIAL: Yes (pulse- or tone-dial)
PHONE: RJ11 (modular) connection

Features. The Intelligent Modem is Bell 103-compatible with operating speeds of 0-300 baud. It incorporates a microprocessor and associated firmware to provide auto-dial and auto-answer capabilities.

The Intelligent Modem has a single modular-phone (RJ11C) socket for connection to an appropriate phone outlet; however, an inexpensive T-adaptor can be used to connect a modular phone. Dialing commands enable either pulse- or touch-tone dialing and automatic redialing. Auto-answer commands can set the modem to answer the phone automatically. A local self-test can be implemented through keyboard commands.

Discussion. The modem features two states: the command mode and the data mode. The modem assumes the command mode when first powered up. One- or two-character commands are entered from the keyboard. The commands are used to set RS-232C parameters, such as baud rate and duplex, and also to implement the various dialing instructions. The data mode is used when data is being transmitted to or received from a remote modem.

Baud rate, escape sequences, sign-on and answer-back characters, and connection messages can be programmed into the modem's memory. Additional commands will put the modem into answer mode or originate mode. The modem can be programmed to be completely transparent in the data mode, ignoring escape codes that would normally cause a disconnection. This transparent operation can be used for transferring non-ASCII files.

An internal self-test verifies the working condition of the microprocessor, the modulator and demodulator circuitry, and analog filters. The test concludes with an analog loopback test, verifying the working condition of the transmitter and receiver circuitry.

1012 Intelligent Modem

BAUD: 1200/300
COMPUTER: Any RS-232C-compatible system
TYPE: Bell 212A/103/113
SOFTWARE: Not included
AUTO ANSWER: Yes
AUTO DIAL: Yes (pulse- or tone-dial)
PHONE: RJ11 (modular) connection

Features. The 1012 Intelligent Modem is Bell 212A- and Bell 103-compatible, with operating speeds of 1200 baud and 0 through 300 baud. It incorporates a microprocessor and associated firmware to provide auto-dial and auto-answer capabilities.

The 1012 has a single modular-phone (RJ11C) socket for connection to an appropriate phone outlet; however, an inexpensive T-adaptor can be used to connect a modular phone. Dialing commands enable either pulse- or touch-tone dialing and automatic redialing. Auto-answer commands can set the modem to answer the phone automatically.

The Intelligent modem can be configured to work with a "dumb" terminal or an "intelligent" computer. A series of nine rocker switches in the back helps to configure the modem for various tests and specific configurations. Tests include an analog loopback test, remote digital loopback test, local digital loopback test, and a self-test.

Discussion. The modem features two states: the command mode and the data mode. The modem assumes the command state when first powered up. One- or two-character commands are entered from the keyboard. The commands are used to set RS-232C parameters, such as baud rate and duplex, and also to implement the various dialing instructions. The data mode is used when data is being transmitted to or received from a remote modem.

A single number can be entered into the modem's memory from a keyboard entry. That number can be redialed, displayed, or changed from single-character entries.

Baud rate, escape sequences, sign-on and answer-back characters, and connection messages can be programmed into the modem's memory. Additional commands will put the modem into answer mode or originate mode. The modem can be programmed to be completely transparent in the data mode, ignoring escape codes that would normally cause a disconnection. This transparent operation can be used for transferring non-ASCII files.

A rocker switch on the back panel implements an analog loopback test. Characters entered through the keyboard of a terminal or computer enter the transmit circuitry of the modem, are fed back through the receive circuitry, and are displayed on the video-display unit for verification.

Another rocker switch on the back panel implements a self-test, generating a pattern of characters that are modulated and demodulated. Any error causes an LED on the front panel to blink.

A digital loopback test causes data transmitted to a remote modem to be looped back to the original modem, to verify the working condition of all equipment, including phone lines.

There is also a rocker switch which designates the modem for use with a terminal, providing English responses to entries, or with a computer, providing only single-digit responses.

PC: IntelliModem

BAUD: 300/1200
COMPUTER: IBM PC/IBM XT
TYPE: Bell 212
SOFTWARE: Included (PC:Intellicom)
AUTO ANSWER: Yes
AUTO DIAL: Yes (pulse- or tone-dial)
PHONE: RJ11 (modular) connection

Features. The PC:IntelliModem is an internal modem that inserts into one of the option slots of the IBM PC or IBM XT. It requires that your computer have at least 64K of memory, at least one diskette drive, the IBM Personal Computer disk operating system and disk BASIC language, and an 80-column monitor (when using the software included).

The modem includes jacks for connecting an RJ11 phone. One jack goes to the modular wall socket, one to the telephone. An optional jack is included for connecting to a separate handset for voice communications.

The modem includes auto-dial features, using either pulse-dialing or tone-dialing. Working with the software included, a directory of numbers can be used to automatically dial a number, initiate a logon sequence, and redial if the number is busy.

The modem also includes an auto-answer feature that allows callers to contact an unattended system and leave messages. The PC:IntelliCom will automatically answer the call, record the data to diskette, hang up the phone, and name the data file.

A special voice mode allows the user to switch conveniently between transmitting data and talking. Executing the voice mode requires a separate handset.

The device includes a menu-driven selection of various modes and features. The device can be used with other communication packages, including Crosstalk XVI.

Discussion. The documentation provides thorough instructions for installing the modem into the IBM PC or IBM XT. A jumper setting on the board itself configures the device for either of the two possible serial ports.

Prior to use, the DOS must be added to the diskette containing PC:IntelliCom; the documentation carefully describes this process. Many of the functions of the modem are transparent when using the accompanying software. A menu-driven format, providing such options as auto-dial and auto-answer, will prompt you for necessary data, such as phone numbers and baud rates. Actually, the modem assumes one of two states: the command mode or a data mode. It enters the command mode when first powered up. In the command mode, it recognizes one- or two-character commands: "C1" activates the auto-answer feature; "SB1" sets the baud rate to 300, and "E1" echoes commands back to the software for display.

The software allows you to press a single key (a menu option); then automatically sends the necessary commands to the modem to execute that feature. For a more thorough discussion on the PC:IntelliCom, see Chapter 9.

The modem includes both a data mode, for transferring data across the phone lines, and a voice mode, for conversation across the same lines. You can establish contact in the data mode, transferring data in the process. Without having to redial, you can switch to voice mode and converse with someone on the other end of the line with the same capabilities, and you can switch back to data mode to continue transferring data.

You can create a dialing directory of up to 99 useful names and numbers; the numbers can have up to 32 digits. Additional commands necessary for logging on to a system, as well as a specific baud rate for that number, can be included in the directory. The logon sequence cannot recognize responses from the other system, but it can include delays for proper timing. The listing can include an alternate number, in case of a busy signal, or in case of no answer within 40 seconds. Executing a call is as easy as entering the row number of the directory and pressing return. Of course, calls can be made manually from the keyboard without using the directory.

The modem can be configured for RS-232C parameters. A default setting can be established for all transmissions. Individual baud rates can be included in the phone directory.

A data-capture mode will automatically save all

data received on your screen. You can leave and enter this mode during transmission to selectively save information. The mode can be left on in the auto-answer mode to save incoming messages.

CERMETEK MICROELECTRONICS, INC.

1308 Borregas Ave.
Sunnyvale, CA 94088
(408) 752-5000

Cermetek, a manufacturer of modems, also makes individual components. The products sell through a number of dealers and distributors.

Cermetek warrants, to the original user, that its products are free from defects in workmanship and material for a period of 1 year from the date of purchase.

Models reviewed and suggested retail prices:

Info-Mate 212A	$595.00
Info-Mate 212PC	$480.00

Info-Mate 212A

BAUD: 300/1200
COMPUTER: Any RS-232C-compatible system
TYPE: Bell 212A/103
SOFTWARE: Not included
AUTO ANSWER: Yes
AUTO DIAL: Yes (pulse-dial)
PHONE: RJ11 (modular) connection

Features. This external modem is compatible with both Bell 212A and Bell 103/113. In the high-speed mode, data transfers at 1200 baud in either a bit-synchronous format or a character-asynchronous format. A standard DB25 connection interfaces with most computers using the RS-232C standard.

The Info-Mate 212A supports a 52-location memory that can store either telephone numbers or logon messages. Batteries protect the loss of data from memory during a loss of power.

Advanced features include a built-in serial auto-dialer, a loopback test (analog, local digital, or remote digital), and an auto-answer mode.

This modem interfaces to the phone system through two standard RJ11C connections—one for the wall outlet and one for the telephone. Power comes from a wall-plug power module, included with the device.

Info-Mate 212A recognizes 16 commands for executing various dialing and answering features. The front-panel display indicates terminal ready, modem ready, send data, receive data, high speed, modem check, and test mode. The additional purchase of communications software is necessary.

Discussion. The various modes of operation become activated through two-character commands, sent either directly from the keyboard of a terminal or a computer or indirectly from the modem's memory. Dialing commands can be programmed to use either pulse- or touch-tone dialing, redial a busy number, dial an alternative number if busy, or hang up after a preset amount of time. A series of commands aid in entering and editing dialing and answering instructions into the modem's memory.

A local analog loopback test verifies the modem's modulation/demodulation functions, as well as the terminal-to-modem connection. Data modulated by the modem, instead of transmitting normally over the phone lines, feeds back into the modem and returns to the terminal. Faulty returned data indicates a problem with your modem, terminal, or software.

A digital loopback test requires the assistance of an operator at the other end of the line. A signal from your computer goes over the phone lines to the receiving modem. That modem, instead of passing on the data to its terminal, feeds the data back into the phone lines and returns it to your end. Problems in the test indicate a problem with your equipment, the phone lines, or the modem at the other end.

The DIP switches at the rear of the modem select various features: asynchronous mode or synchronous mode; seven-, eight-, or nine-bit characters; override control of the CTS, DSR, DCD, and DTR lines; exclusion-key telephone control; and analog or digital loopback tests.

Info-Mate 212PC

BAUD: 300/1200
COMPUTER: IBM PC, XT, or compatible
TYPE: Bell 212A/103
SOFTWARE: Included (Modem-Mate)
AUTO ANSWER: Yes
AUTO DIAL: Yes (pulse- or tone-dial)
PHONE: RJ11 (modular) connection

Features. This internal modem is compatible with both Bell 212A and Bell 103/113. In the high-speed mode, data transfers at 1200 baud, in either a bit-synchronous or character-asynchronous format. The modem inserts into any of the empty slots inside the IBM computer. An auxiliary RS-232C port is provided for connecting alternative RS-232C devices.

Advanced features include a built-in serial auto-dialer, a loopback test (analog or digital, local or remote), and an auto-answer mode. This modem interfaces to the phone system through two standard RJ11C connections—one for the wall outlet and one for the telephone.

The modem includes the telecommunications program Modem-Mate and provides a discount on the separate purchase of Crosstalk XVI from Microstuf.

Discussion. The various modes of operation are implemented through one- and two-character commands, sent directly from the keyboard of a terminal or a computer. InfoMate 212PC automatically sets its speed and parity to match your computer.

Dialing commands can be programmed to use either pulse- or touch-tone dialing, redial a busy number, dial an alternative number if busy, or hang up after a preset amount of time.

A local-analog loopback test verifies the modem's modulation/demodulation functions as well as the terminal-to-modem connection. Data modulated by the modem, instead of transmitting normally over the phone lines, feeds back into the modem and returns to the terminal. Faulty returned data indicates a problem with your modem, terminal, or software.

A digital loopback test requires the assistance of an operator at the other end of the line. A signal from your computer goes over the phone lines to the receiving modem. That modem, instead of passing on the data to its terminal, feeds the data back into the phone lines and returns it to your end. Faulty returned data indicates a problem with your equipment, the phone lines, or the modem at the other end.

A series of option switches on the modem addresses the communications channel; selects word length; controls CTS, DSR, and DCD lines; selects the auxiliary RS-232C channel, and selects the type of telephone.

COMMODORE BUSINESS MACHINES INC.

1200 Wilson Dr.
West Chester, PA 19380
(215) 431-9100

Commodore is one of the leaders in the low-cost home-computer market. The company manufactures a number of peripherals and markets software for its two most successful computers: the VIC 20 and the Commodore 64.

Commodore provides a 90-day warranty on its products. We have had good experience with Commodore concerning the repair of their equipment after the warranty has expired. Generally, the product can be returned to one of their authorized service centers and, for a reasonable charge, repaired or replaced on the spot.

Models reviewed and suggested retail prices:

VICMODEM (1600)	$ 69.95
Automodem (1650)	$149.95

VICMODEM (1600)

BAUD: 300
COMPUTER: Commodore 64, VIC 20
TYPE: Bell 103
SOFTWARE: Included
AUTO ANSWER: No
AUTO DIAL: No
PHONE: RJ11 (modular) connection

Features. This Bell 103-compatible modem inserts directly into the user port of the Commodore 64 or VIC 20. It connects to the telephone through the cable that normally connects to the handset of a modular phone. The modem comes with the necessary communications software. A single switch sets answer and originate modes. A single LED indicates a carrier-detect signal.

The modem is powered from the computer and requires no additional power supplies. After a call is dialed from the phone, the cable to the handset of a modular phone must be disconnected and reinserted into the modem for data communications.

The VICMODEM requires a modular (RJ11C) phone with a full keystroke number pad. Most phones today use either of two types of keypad: short-stroke or full-stroke. The short-stroke keys depress 1/32 inch before coming to rest. The full-stroke keys depress ⅛ inch. Differences in voltages provided by the two types of phones make the VICMODEM unsuitable for phones with short keystroke.

Discussion. Included with the modem and software are free memberships to CompuServe and the Dow Jones News/Retrieval Service. The product has, however, some distinct limitations. The procedure for establishing contact with a remote computer includes dialing by hand the number of the service, disconnecting the cable to the handset of the phone, and reconnecting the cable to the modem; the dismantling of the telephone each time can be inconvenient, and the modem hangs precariously from the user port, presenting the risk of disconnection.

Recent versions of the VICMODEM include software for the Commodore 64, but this lacks downloading and uploading capabilities. The software also lacks auto-log features that would assist in logging on to a remote system.

More sophisticated software such as Superterm, from Midwest Micro, or Vidtex, from CompuServe, can be purchased separately and substituted for the program included. The documentation provides considerable background in telecommunications and necessary information for installation and use.

Automodem (1650)

BAUD: 300
COMPUTER: Commodore 64, VIC 20
TYPE: Bell 103
SOFTWARE: Included
AUTO ANSWER: Yes
AUTO DIAL: Yes
PHONE: RJ11 (modular) connection

Features. The Automodem inserts physically into the user port of the Commodore 64 and simultaneously connects to a modular phone. A single switch on the modem allows you to conveniently select between normal phone use and data communications. The modem features automatic answering and automatic dialing. It includes software for basic telecommunications, including auto-dial and auto-answer modes.

The Automodem eliminates the need for an RS-232C interface. A built-in phone adapter allows you to use the modem with any modular-type phone. The cable that normally connects the wall socket to the phone inserts into the modem. A short cable is included that connects the modem to the phone. A T-adaptor, normally required for this type of connection, is built into the modem.

The modem includes the program Term 64. No additional software is required to use the Automodem for telecommunications.

Discussion. The Automodem comes complete with software and requires no additional hardware for interfacing. Settings on the modem allow you to switch between phone mode, for making regular phone calls, and data mode, for exchanging data with your computer. An additional setting selects either full-duplex or half-duplex and must correspond to a similar setting made through software commands. A third switch places the modem in either originate or answer mode. With the modem inserted in the user port, the connection is vulnerable to movement of the keyboard.

The program Term 64 is included with the modem and provides the essentials for telecommunications. Information can be transferred between two computers. Up to ten phone numbers can be entered through the keyboard, saved to tape or diskette, and automatically dialed upon your request. If you wish to dial a number not on the list, you can enter the number through the keyboard and press return; the modem dials the number automatically.

The auto-dial feature has one limitation: alternative long-distance systems which require touch-tone dialing must be dialed manually.

The software included cannot download or upload from diskette. Other communications software, which include those features, can be purchased separately and used with the modem as an alternative to Term 64.

With Term 64 running and the modem in the data and answer modes, the switched-on computer can respond automatically to the computer originating a call. The modem will answer the call automatically and display any incoming information.

COMMUNICATIONS AND PERIPHERALS MARKETING CORPORATION

2318-B Walsh Ave.
Santa Clara, CA 95051
(408) 986-1555

CPM is currently marketing a variety of external modems and telecommunications equipment. Each modem comes with a 1-year warranty. If you have any problems, you can send the product to their Santa Clara center, which will repair and return it. After the warranty expires, repairs or replacement can still be made through the same address.

Models reviewed . 'd suggested retail prices:

MM-102	$ 64.95
Automodem 1200A	$279.95

MM-102

BAUD: 300
COMPUTER: Any RS-232C-compatible system
TYPE: Bell 103
SOFTWARE: Not included
AUTO ANSWER: No
AUTO DIAL: No
PHONE: RJ11 (modular) connection

Features. This modem is compatible with Bell 103 and can send and receive at rates of 0-300 baud. It connects to the wall socket of a modular phone and provides an additional socket to connect the telephone. A three-position switch selects between originate, answer, and phone modes. In the phone mode, the telephone can be used for normal conversations. In the originate and answer modes, your computer links through the modem to the telephone lines for the transmission of data. A carrier-detect light indicates a proper connection with a modem at the other end.

Discussion. The MM-102 connects with any RS-232C-compatible system. The modem has a socket which accepts a standard DB25 plug for connection to an interface. It requires an RJ11C (modular-phone type) wall socket and provides an outlet and cable for connecting a RJ11C telephone. Included is a modular-power supply, which plugs into a standard 120-volt ac power outlet. Operation of the modem requires the additional purchase of communications software which uses the standard RS-232C protocol.

The modem can be connected simultaneously to both your phone and your computer. A single switch selects between the use of the phone for conversation and for data communications. In the data mode, a carrier-detect light indicates a proper connection with another modem.

The modem has no DIP switches; it sets itself automatically to work with the RS-232C parameters established through software. The modem has a frequency-attenuation feature that, in either the answer or originate modes, eliminates noise from the telephone. After establishing contact with a host computer, inadvertently picking up the handset will not interfere with data transmissions.

Auto Modem 1200A

BAUD: 300/1200
COMPUTER: Any RS-232C-compatible system

TYPE: Bell 103/Bell 212A
SOFTWARE: Not included
AUTO ANSWER: Yes
AUTO DIAL: Yes (pulse-dial)
PHONE: RJ11 (modular) connection

Features. This Bell 212A-compatible modem can transfer data at 1200 baud, in either an asynchronous or synchronous mode. It can also transfer at 0 to 300 baud in a Bell 103/113 format.

Advanced features include a built-in serial auto pulse-dialer, loopback tests (analog or digital, local or remote), and an auto-answer mode.

This modem connects to the phone system through a standard RJ11C connection. Power comes from a wall-plug power module, included with the device. The additional purchase of communications software is necessary.

Discussion. The various modes of operation are selected manually through three switches on the front panel and DIP switches on the back. A "Talk/Data" switch on the front panel alternates the use of the phone for either normal conversation or for data transfer. A second switch selects between 1200 baud and 300 baud. A third switch simply powers up the modem.

A call can be initiated manually by simply dialing the number and, after establishing verbal communications, changing the switch on the front panel to "Data." Data can then be transferred between the two modems. The auto-dial feature can be implemented by entering, through the keyboard, the letter "D" and the phone number. If a connection does not occur within 18 seconds, the modem will disengage the line and display "DISCONNECT." The letter "R" will automatically redial the previous number. The auto-answer feature can be used to automatically answer an incoming call.

A local analog loopback test verifies the working condition of the modem's modulation/demodulation functions, as well as the terminal-to-modem connection. Data modulated by the modem, instead of transmitting normally over the phone lines, feeds back into the modem and returns to the terminal. Faulty returned data indicates a problem with your modem, terminal, or software.

A digital loopback test needs the assistance of an operator at the other end of the line. In this test, a signal from your computer is transmitted over the phone lines to the receiving modem. That modem, instead of passing on the data to its terminal, feeds the data back into the phone and returns it to your end. Faulty returned data indicates a problem with your equipment, the phone lines, or the modem at the other end.

The DIP switches at the rear of the modem select the asynchronous or synchronous modes, seven- or eight-bit characters, and analog or digital loopback tests. A series of eight LEDs on the front panel indicate the current status of the modem.

COMPUTER COMMUNICATIONS SPECIALISTS, INC.

Atlanta/Peachtree Corners Business Park
6683 Jimmy Carter Blvd.
Norcross, GA 30071
(404) 441-3114

In addition to the Audiomodem, Computer Communications Specialists market software for the IBM Series I and the IBM PC. The Audiomodem is warranted for 90 days. A service agreement covering malfunctions occurring after the 90 days can be purchased separately.

The model reviewed and suggested retail price:

Audiomodem	$4895.00

Audiomodem

BAUD: 0-1200
COMPUTER: Any RS-232C-compatible system
TYPE: Bell 202S
SOFTWARE: Included
AUTO ANSWER: Yes
AUTO DIAL: Yes (pulse- or tone-dial)
PHONE: RJ11 (modular) connection

Features. The Audiomodem is a Bell 202S-compatible modem capable of responding with synthesized speech to inputs from touch-tone telephones or virtually any terminal device. The modem features auto-dial, using pulse- or touch-tone dialing. It connects directly to a telephone network through RJ11C jacks and provides half-duplex operation up to 1350 baud over public telephone lines. The Audiomodem connects to computers through a standard RS-232C interface.

The Audiomodem comes with a standard set of approximately 100 words and phrases. Custom words and phrases are also available through Computer Communications Specialists.

Four DIP switches, located on the main circuit board, select between four modes of operation: a stand-alone receiver mode, for when the modem connects to a stand-alone receiver, such as an MSI or a Telxon unit; a diagnostic mode, for verifying the working condition of the modem; a computer-connect mode, for when the modem connects to a host computer capable of sending it commands; and a burn-in mode, used, prior to shipment, by the manufacturer.

Discussion. The Audiomodem operates in a half-duplex Bell 202S format. It was designed for the order-entry types of applications. For example, a caller might hear from the modem a message such as: "Thank you for calling Computer Communications Specialists. At the sound of the tone please send your order." The Audiomodem then functions in a normal half-duplex mode, inputing data from the phone lines, converting it to digital forms, and transmitting it to the computer. After receiving and verifying the data, the modem makes an appropriate verbal response, such as: "We appreciate your order. Your confirmation number is 1234."

The Audiomodem can also be configured to recognize touch-tone entries. It answers the phone and speaks an introductory message. It then monitors the phone input for touch tones. A caller can place orders by simply pressing numbers.

The Audiomodem receives directions through a series of commands from the host computer. A ring-enable command activates the ring indicator signal to the computer when a call is detected. A pass-thru command causes the modem to answer an incoming call and give the answer-back tone. The modem then monitors the line, and, after receiving a proper carrier signal, it begins passing data to the computer. A speak command causes the modem to answer the phone, give an answer-back tone, and then speak a programmed message. After giving the message, the modem then monitors the line for a valid carrier tone. An intermediate speak command allows the modem to monitor the DTR line to the host computer and, after receiving data, provide verbal messages. A touch-tone command causes the modem to answer the phone and speak a message. After giving the message, it monitors the line for touch-tones and saves the corresponding numbers in memory. An auto-dial command causes the modem to attempt the touch-tone dialing of a number. If the call is unsuccessful, it will attempt a pulse-dial. Upon making the connection and detecting a carrier tone, the modem automatically begins transferring data. If it does not detect a carrier tone, it can issue a programmed verbal message.

The modem includes a set of thorough diagnostics to check the hardware and firmware statuses. The modem also recognizes a check-sum command that will run a check sum on a particular block of RAM to ensure data integrity. For the RAM, The Audiomodem includes a 10-year battery.

CONCORD DATA SYSTEMS

303 Bear Hill Rd.
Waltham, MA 02154
(617) 890-1394

Concord Data Systems designs, manufactures, and markets high technology communications equipment for data transmission over telephone lines and broadband cable. The company offers a family of international and domestic 2400-baud full-duplex modems. Recently, Tymnet, Inc., one of the nation's largest public-data networks, began installing Concord modems to expand its low-cost 2400-baud asynchronous dial-up service.

Concord warrants its products to perform in accordance with its published specifications for a period of 12 months from the date of shipment.

Models reviewed and suggested retail prices:

224 Data Modem	$ 995.00
224 Autodial Modem	$1195.00

224 DATA Modem

BAUD: 2400/1200 asynchronous/ synchronous
COMPUTER: Computer with RS-232C, or CCITT V.24/V.28 interface
TYPE: Bell 212
SOFTWARE: Not included
AUTO ANSWER: Yes
AUTO DIAL: No
PHONE: RJ11 (modular) connection

Features. The 224 Data Modem is an external modem that connects through a standard DB25 cable using RS-232C protocol. It can function as a 2400-baud full-duplex modem for operation on two-wire switched-network lines. It contains an adaptive equalizer which automatically adjusts the modem's receiver to enhance performance, even on marginal lines. It operates in both synchronous and asynchronous modes at speeds of 2400 and 1200 baud.

The modem includes jacks for RJ11 connection to phone lines. One jack goes to the modular wall socket; an optional jack is included for connecting to a separate handset.

Front-panel switches allow the alternate use of phone lines for data or voice transmission. A second pair of switches selects between answer and originate modes of transmission. Additional panel settings enable auto answer, data fallback, local digital or remote digital loopback tests, local analog loopback test, an internal pattern generator, and error detection. Additional software is required for interaction with specific computers.

Discussion. The 224 Data Modem comes as either a self-contained desk-top modem or a board for inserting into a rack assembly. The desk-top model requires a nearby power supply.

The modem includes both a data mode, for transferring data across the phone lines, and a voice mode, for conversation across the same line. Most of the various configurations and modes of operation are instituted with a series of DIP switches and straps located within the modem housing. Switch settings can enable an automatic-abort process if the carrier signal becomes too weak. Settings configure the device as synchronous or asynchronous. An optional model of the 224 Data Modem will provide that switch on the front panel.

Timing controls and some of the loopback tests are also implemented from DIP-switch settings. A DTR signal can be forced on through switch settings, and the DCD signal looped to RTS signal for some applications.

The modem can handle two basic loopback tests. An analog loopback test verifies that your modem and computer are functioning properly. This test returns characters, entered at the keyboard, just before they reach the telephone line. Additionally, the front panel can display a test pattern; flashing LEDs indicate errors.

A digital loopback test checks remote equipment. The modem loops back signals received from a remote modem. A command can also be sent that will cause the remote modem to execute a digital loopback at the remote site.

224 Autodial Modem

BAUD: 2400/1200 asynchronous/ synchronous
COMPUTER: Computer with RS-232C, or CCITT V.24/V.28 interface
TYPE: Bell 212
SOFTWARE: Not included
AUTO ANSWER: Yes
AUTO DIAL: Yes (pulse- or tone-dial)
PHONE: RJ11 (modular) connection

Features. The 224 Data Modem is an external modem that connects through a standard DB25 cable using RS-232C protocol (Fig. 8-5). It can function as a 2400-baud full-duplex modem for operation on two-wire switched-network lines. It contains an adaptive equalizer which automatically adjusts the modem's receiver to enhance performance even on marginal lines. It operates in both synchronous and

Fig. 8-5. The 224 Autodial modem. Courtesy Concord Data Systems.

asynchronous modes at speeds of 2400 and 1200 baud.

The modem includes jacks for RJ11 connection to phone lines. One goes to the modular wall socket. An optional jack is included for connecting to a separate handset.

Front-panel switches allow the alternate use of phone lines for data or voice transmission. A second pair of switches select between answer and originate modes of transmission. Additional panel settings enable scynchronous or asynchronous, auto-answer, data fallback, local digital or remote digital loopback tests, local analog test, an internal pattern generator, and error detection.

A "keyboard command" mode allows you to enter commands from your terminal to dial and redial phone numbers, configure modems parameters, and initiate testing.

Discussion. The 224 Data Modem comes as either a self-contained desk-top modem or a board for inserting into a rack assembly. The desk-top model requires a nearby power supply.

The modem includes both a data mode, for transferring data across the phone lines, and a voice mode, for conversation across the same line. The auto-dial features are implemented with single-character entries within the "keyboard command" mode. The character "D" followed by a number will automatically dial that number. Numbers can be programmed to redial from 0 to 99 times every 60 seconds. Additional command entries select between pulse- or tone-dialing. The modem can be programmed to call subsequent numbers after hearing a dial tone. Pauses can be inserted between commands. Parity settings and loopback tests can also be executed through "keyboard commands."

Most of the various configurations and modes of operation are instituted with a series of DIP switches and straps located within the modem housing. Switch settings can enable an automatic-abort process if the carrier signal becomes too weak. Timing controls and some of the loopback tests are also implemented from DIP-switch settings. A DTR signal can be forced on through switch settings, and the DCD signal looped to RTS signal for some applications.

The modem can handle two basic loopback tests. An analog loopback test verifies that your modem and computer are functioning properly. The test returns characters, entered at the keyboard, just before they reach the telephone line. Additionally, the front panel can display a test pattern; flashing LEDs indicate errors.

A digital loopback test checks remote equipment. The modem loops back signals received from a remote modem. A command can also be sent that will cause the remote modem to execute a digital loopback at the remote site.

DATASOLVERS, INC.

509 S. Fern
Wichita, KS 67213
(316) 264-5068

Datasolvers produces both 300-baud and 1200-baud internal modems for the KAYPRO computers. Its products are warranted for 90 days against defects in parts and workmanship.

Model reviewed and suggested retail price:

DS1200SMT $427.50

DS1200SMT

BAUD: 1200/300
COMPUTER: KAYPRO
TYPE: Bell 212A
SOFTWARE: Included
AUTO ANSWER: Yes
AUTO DIAL: Yes (pulse- or tone-dial)
PHONE: RJ11 (modular) connection

Features. The DS1200SMT is a Bell 212A-compatible internal modem available for KAYPRO 2, 4, and 10 (Fig. 8-6). It incorporates Hayes-compatible commands for auto dial and auto answer. Detailed instructions are provided for installing the modem into the computer. Two RJ11C ports provide connections to an appropriate wall socket and telephone.

One- and two-character commands, entered

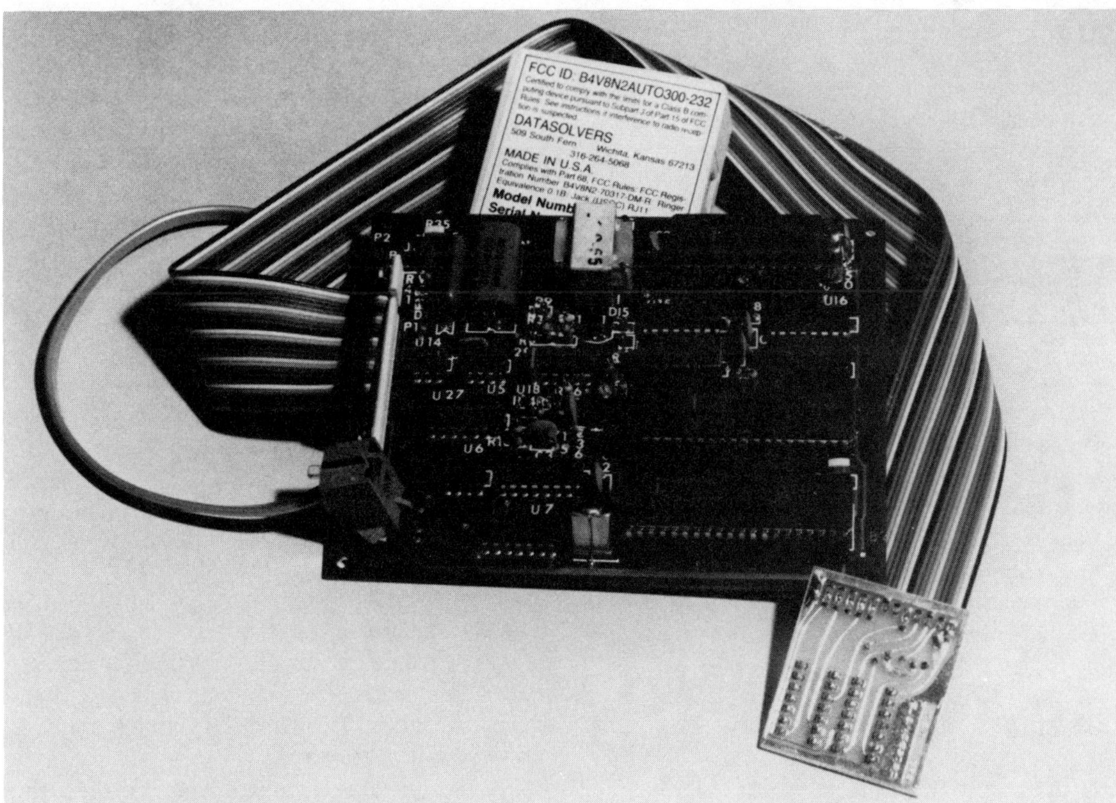

Fig. 8-6. The DS1200SMT internal modem for KAYPRO computers. Courtesy Datasolvers, Inc.

from the keyboard, execute various features on the modem. Commands can be combined, up to 60 characters in length, and entered with a single carriage return. The results of various commands can be displayed on the screen. Results include OK, CONNECT, RING, NO CARRIER, ERROR, and CONNECT 1200.

Discussion. The modem includes the software SMODEMK; however, most communications software compatible with the Hayes command set should work as well.

The modem adjusts itself automatically to accommodate the speed of your computer up to 1200 baud. When dialing a number, it will automatically disconnect if the call goes unanswered after the programmed amount of time. The default time is 30 seconds.

In the auto-answer mode, the modem will automatically answer any incoming call. It will wait up to 17 seconds for a carrier frequency from the caller's computer. If no carrier is present, the modem will automatically disconnect.

Internal switches can be set on the modem to accommodate 10 or 11 bits per character (including start and stop bits).

GANDALF DATA, INC.

1019 South Noel St.
Wheeling, IL 60090
(312) 541-6060

Gandalf Data is a subsidiary of Gandalf Technologies, Inc., one of Canada's largest manufacturers and suppliers of digital data-communications equipment, which also has subsidiaries in Canada and the United Kingdom. Gandalf warrants its modems to be free from defects in parts and workmanship for 1 year.

Model reviewed and suggested retail price:
SAM 212A $618.00

SAM 212A

BAUD: 300/1200
COMPUTER: Any RS-232C-compatible system
TYPE: Bell 103/Bell 212A
SOFTWARE: Not included
AUTO ANSWER: Yes
AUTO DIAL: Yes (pulse- or tone-dial)
PHONE: RJ11 (modular) connection

Features. This external Bell 212A-format modem will also operate in Bell 103/113 format (Fig. 8-7). In the high-speed mode, data transfers at 1200 baud, either character asynchronously or bit asynchronously. The SAM 212A automatically adapts its speed and parity to match your terminal or computer. A standard DB25 connection interfaces with most computers using the RS-232C standard.

The SAM supports a 52-location memory that can store either telephone numbers or logon messages. Batteries protect the loss of data from memory during a loss of power. Advanced features include a built-in serial auto dialer, an analog and digital (local or remote) loopback test, and an auto-answer mode.

This modem interfaces to the phone system through two standard RJ11C connections, one for the wall outlet, and one for the telephone. Power is provided from a wall-plug power module included with the device.

SAM recognizes 16 commands for executing various dialing and answering features. The front-panel display indicates terminal-ready, modem-ready, send-data, receive-data, high-speed, modem-check, and test modes.

Discussion. The various modes of operation are implemented through two-character commands, sent directly from the keyboard of a terminal or a computer or indirectly from the modem's memory. Dialing commands can be programmed to use pulse or touch-tone dialing, redial a busy number, dial an alternative number if busy, or hang up after a preset amount of time. A series of commands aids in entering and editing dialing and answering intructions into the modem's memory.

A local analog loopback test verifies the

Fig. 8-7. The SAM 212A external modem. Courtesy Gandalf Data, Inc.

modem's modulation/demodulation functions, as well as the terminal-to-modem connection. Data modulated from the modem, instead of transmitting normally over the phone lines, feeds back into the modem and returns to the terminal. Faulty returned data indicates a problem with your modem, terminal, or software.

A digital loopback test requires the assistance of an operator at the other end of the line. A signal from your computer transfers over the phone lines to the receiving modem. That modem, instead of transferring the data to its terminal, feeds the data back into the phone lines and returns it to your end. Faulty returned data indicates a problem with your equipment, the phone lines, or the modem at the other end.

The DIP switches at the rear of the modem select between various options: asynchronous or synchronous modes; seven-, eight-, or nine-bit characters; override control of the CTS, DSR, DCD, and DTR lines, exclusion-key telephone control; and analog or digital loopback tests.

GENERAL DATACOMM, INC.

One Kennedy Ave.
Danbury, CT 06810
(203) 797-0711

General Datacomm is a major manufacturer of modems and communications equipment. In addition to the modems reviewed here, the company has a variety of high-speed synchronous sets for business and commercial applications. General Datacomm warrants its products to be free from defects in materials and workmanship for 1 year.

Models reviewed and suggested retail prices:

201C	$975.00
212A/L	$675.00
212A/ED	$575.00

201C

BAUD: 2400/1200 synchronous
COMPUTER: Any RS-232C-compatible system

TYPE: Bell 201C CCITT V.26
SOFTWARE: Not included
AUTO ANSWER: No
AUTO DIAL: No
PHONE: RJ11 (modular) connection or four-wire private line

Features. The 201C operates at 2400 baud full duplex over unconditioned four-wire voice-grade circuits or half-duplex over two-wire switched networks. It comes as either a stand-alone unit with its own enclosure or as a card to be inserted in a rack mount (Fig. 8-8).

The modem is shipped from the factory with field-selectable options preselected for configurations requested by the user. Selectable options include 1200- or 2400-baud synchronous transmission, variable carrier-detect sensitivities, automatic answering, two-wire or four-wire phone connection, and private line or switched network.

A built-in quality-monitor circuit provides a continuous performance check of the received carrier signal. Numerous built-in diagnostics and self-tests allow digital and analog loopback tests.

Discussion. Front-panel switches execute self-tests, remote test, digital loopback, analog loopback, and test patterns, Front-panel lights indicate the data-terminal ready, modem ready, request to send, clear to send, carrier on, as well as data-transmit and data-receive lines. Additional LEDs are used for diagnostics and self-tests.

Numerous field-selectable options are configured through a series of DIP switches and jumper plugs located inside the modem. The list of options include the following:

- Two-wire or four-wire operation.
- Controlled or continuous carrier.
- Request-to-send/clear-to-send delays of 7.1, 35, 150, or 690 ms, or controlled/continuous clear-to-send.
- Control of data-set-ready by an 828 or 829 data-auxiliary set.
- Satellite timing intervals of 53 or 275 ms.
- Transmit-line signal-level variation of 0 to — 15 dBm in 1 dBm increments.
- 600-ohm or 900-ohm receive-line impedance.
- 600- or 900-ohm transmitter impedance.
- Delay and amplitude equalization.
- Carrier-detect sensitivity ranges of −24 to −29 dBm, −34 to −39 dBm, or −44 to −29 dBm.
- Internal-, external-, or receiver-loopback transmit timing.
- New sync via business equipment interface pin.

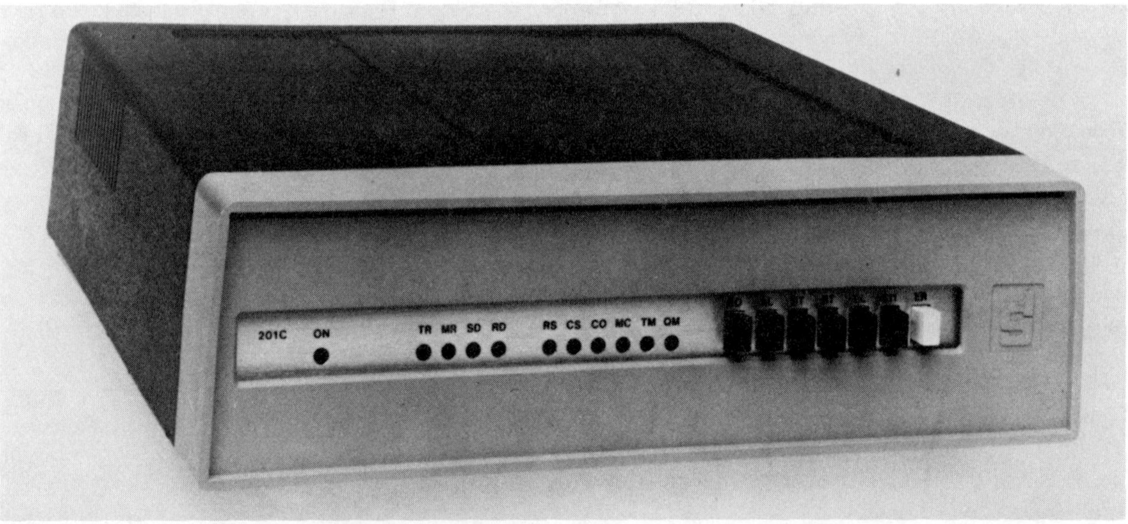

Fig. 8-8. The 201C modem. Courtesy General Datacomm, Inc.

Fig. 8-9. The 212A/L modem. Courtesy General Datacomm, Inc.

- Automatic or manual answering.
- Common or separate signal and frame grounds.
- Data-set-ready indication during analog loopback.
- Receiver-signal element-clock enable when carrier detect is off.
- External-analog loopback control or receiver digit clock on business-equipment interface pin 18.
- Scrambler/descrambler enable.
- Anti-streaming and abort timer.
- Anti-streaming timer intervals of 3 to 27 seconds.
- Programmable or permissive transmit-line signal level.
- Make-busy during analog loopback.
- Grounded- or floating-contact second-pair make-busy.
- B or C answer-tone sequence.
- Quality-monitor output available on business-equipment interface pin 21.
- Data-rate control via business-equipment interface pin.
- Data rate of 1200, 2000, or 2400 bps.
- Delayed analog loopback.
- Data-set-ready indication during self-test, remote test, and digital loopback.
- Analog loopback control via business-equipment interface pin.
- Receiver inhibit during transmissions.
- Receiver dibit-clock enable when carrier-detect is off.
- Control of request-to-send/clear-to-send delay via telephone-connector pin.
- Quality-monitor inhibiting of carrier-detect.
- LED or incandescent lamp auxiliary-telephone-set lights.
- RTC/502 or 5654 auxiliary telephones.

212A/L

BAUD: 1200 synchronous/asynchronous, 300 asynchronous
COMPUTER: Any RS-232C-compatible system
TYPE: Bell 103/113/212A
SOFTWARE: Not included
AUTO ANSWER: Yes
AUTO DIAL: No
PHONE: RJ11 (modular) connection

Features. The 212A provides full-duplex data communications for switched network and private lines, operating asynchronously at 0 to 300 bps with 103 or 113 modems, and synchronously or asynchronously at 1200 bps with 212A modems (Fig. 8-9).

Calls are made manually and answered either manually or automatically. The unit comes as a stand-alone or as a rack-mount shelf.

A special security feature isolates the modem from the telephone line for a short time after a disconnect to allow the computer time to logout before receiving another call. This feature prevents a second caller from accidently accessing a confidential data base.

The modem includes a built-in brown-out protection, enabling it to operate with low-power line voltages and a comprehensive set of internal diagnostics. A digital-adaptive equalizer corrects received signals for a broad range of line degradations to enhance performance over switched networks.

Discussion. A series of five switches on the front panel executes various diagnostics, including a self-test, remote digital loopback, local digital loopback, and analog loopback. An additional panel switch selects between high-speed or low-speed operation. Front panel LEDs include a make-busy-mode light, terminal-ready light, data-sending light, data-receiving light, high-speed indicator, and test mode.

Field-selectable options are configured through a series of DIP switches and internal straps. The configurable options are as follows:

- Asynchronous operation at 0-300 bps and 1200 bps, or synchronous operation at 1200 bps.
- Character lengths of 8, 9, 10 or 11 bits.
- Receive-space disconnect.
- Send-space disconnect.
- Loss-of-carrier disconnect.
- High-speed or dual-speed operation.
- Modem-speed indication (CI) or DTE.
- Clear-to-send and receive-line-signal detector, common or separate.
- Answer-mode indication.
- Common or separate signal and frame grounds.
- Data-set-ready indication during analog-loopback test.
- Test-mode indication or DTE.
- Make-busy/analog loopback control via front panel and/or DTE.
- Remote digital loopback.
- Received-data inhibit during modem hek.
- Internal, external, or slave timing.
- 502/RTC or 565/2565 auxiliary telephone with incandescent or LED lights.
- Transmit-line signal-level variation of 0 to –15 dBm in 1 dBm increments.
- Permissive or programmable transmit-line signal level. Floating- or grounded-contact second pari make-busy.
- 600- or 900-ohm impedance.
- Continuous ringing or busy signal on telephone line while modem out of service.
- Originate-answer, originate-only, or answer-only.

212A/ED

BAUD: 1200 Synchronous /asynchronous, 300 asynchronous
COMPUTER: Any RS-232C-compatible system
TYPE: Bell 103/113/212A
SOFTWARE: Not included
AUTO ANSWER: Yes
AUTO DIAL: Yes (pulse- or tone-dial)
PHONE: RJ11 (modular) connection

Features. The 212A provides full-duplex data communications for switched network and private lines, operating asynchronously at 0 to 300 bps with 103 or 113 modems, and synchronously or asynchronously at 1200 bps with 212A modems (Fig. 8-10).

An automatic dialer enables a call to be originated from the terminal through the modem's front panel buttons, or manually through the phone. Answering calls can also be performed automatically or manually.

Discussion. A set of two-stroke commands executes auto-dial instructions from the keyboard of a terminal. Phone numbers can be stored in the modem and dialed automatically from five buttons on the front panel. A blind-dialing feature allows the modem to dial a number in spite of a weak or absent dial tone. An automatic linking feature will automatically dial another number if the first is busy. A redial

Fig. 8-10. The 212A/ED modem. Courtesy General Datacomm, Inc.

feature will automatically redial the number a specified number of times. A small speaker within the modem permits you to monitor calls and hear dial tones, busy signals, etc.

Field-selectable options are configured through a series of DIP switches and internal straps. The configurable options are as follows:

- Character lengths of 8, 9, 10, or 11 bits.
- Receive space disconnect.
- Send space disconnect.
- Loss-of-carrier disconnect.
- High-speed or dual-speed operation.
- Modem-speed indication (CI) or DTE.
- Clear-to-send and receive-line-signal detector, common or separate.
- Answer-mode indication.
- Common or separate signal and frame grounds.
- Data-set-ready indication during analog-loopback test.
- Test-mode indication to DTE.
- Make-busy/analog-loopback control via front panel and/or DTE.
- Remote digital loopback.
- Received-data inhibit during modem hek.
- Internal, external, or slave timing.
- 502/RTC or 565/2565 auxiliary telephone with incandescent or LED lights.
- Transmit-line signal-level variations of 0 to –15 dBm in 1 dBm increments.
- Permissive or programmable transmit-line signal level.
- Floating- or grounded-contact second pari make-busy.
- 600- or 900-ohm impedance.
- Continuous ringing or busy signal on telephone line while modem out of service.
- Originate-answer, originate-only, or answer-only operation.

HAYES MICROCOMPUTER PRODUCTS, INC.

5923 Peachtree Industrial Blvd.
Norcross, GA 30092
(404) 441-1617

Hayes was an early leader in modem manufacture. Its command set for executing auto-dial and auto-answer features has become a standard in the industry. Hayes produces both hardware and

software products for the microcomputer. Hayes offers a 2-year limited warranty on its hardware products.

Models reviewed and suggested retail prices:

Smartmodem 300	$289.00
Smartmodem 1200	$699.00
Micromodem IIE	$329.00

Smartmodem 300

BAUD: 300
COMPUTER: Any RS-232C-compatible system
TYPE: Bell 103
SOFTWARE: Not included
AUTO ANSWER: Yes
AUTO DIAL: Yes (pulse- or tone-dial)
PHONE: RJ11 (modular) connection

Features. This external 300-baud modem has a female DB25 socket for connection to any computer capable of RS-232C protocol (Fig. 8-11). The back of the modem includes a single RJ11C connector, a power switch, a volume control for the internal speaker, and a connection for the power adapter.

The modem can dial numbers from keyboard entries through touch-tone or pulse-dial environments, wait for a second dial tone (as with PBX systems), and redial the previous number. The small internal speaker allows you to monitor the call, listening for busy signals and wrong numbers. In the auto-answer mode, the modem will answer an incoming call after a programmed number of rings.

The Smartmodem connects to the phone system through a standard RJ11C connector. A special T-adaptor (not included) can be used to provide an additional connection for a telephone.

Discussion. The Smartmodem operates in two functional states: local command and on line. In the local-command mode, ASCII characters have special meanings. For example, the character "D" followed by a number instructs the modem to dial

Fig. 8-11. The Smartmodem 300. Courtesy Hayes Microcomputer Products, Inc.

Fig. 8-12. The Smartmodem 1200. Courtesy Hayes Microcomputer Products, Inc.

that number. In the command mode, entries from the keyboard travel only as far as the modem; they do not transmit over the phone lines to another computer.

The Smartmodem goes into the on-line mode after answering a call or dialing a call and connecting with a distant modem. In the on-line mode, characters transfer normally through the phone lines to a receiving system. For example, the character "D" no longer causes the modem to dial a number, but transfers as the letter itself.

Seven LEDs on the front panel of the Smartmodem indicate modem-ready, terminal-ready, send-data, receive-data, off-hook, carrier-detect, and auto-answer modes.

Eight DIP switches, accessed behind the front panel of the modem, provide some manual programmability. One switch makes available English responses to command codes. For example, the terminal displays CONNECT after the dialing of a number and the making of a connection with a distant modem. Other settings provide for auto answer on first ring, locally echoed characters and result codes, single-line connection to an RJ11 phone, and an enabling mode for commands.

Smartmodem 1200

BAUD: 1200/300
COMPUTER: Any RS-232C-compatible system
TYPE: Bell 212A/103
SOFTWARE: Not included
AUTO ANSWER: Yes
AUTO DIAL: Yes (pulse- or tone-dial)
PHONE: RJ11 (modular) connection

Features. This external 1200-baud modem has a female DB25 socket for connection to any computer capable of RS-232C protocol (Fig. 8-12). It operates in a Bell 212A format, at 1200 baud, and in a Bell 103 format, at 0-300 baud. The back of the modem includes a single RJ11C connector, a power switch, a volume control for the internal speaker, and a connection for the power adapter.

From keyboard entries, the modem can use tones or pulses to dial numbers, wait for a second dial tone (as with PBX systems), and redial the previous number. The small internal speaker allows you to monitor the call, listening for busy signals and wrong numbers.

In the auto-answer mode, the modem will answer an incoming call after a programmed number of rings.

The Smartmodem connects to the phone system through a standard RJ11C connector. A special T-adaptor (not included) can be used to provide an additional connection for a telephone.

Discussion. The Smartmodem operates in two funtional modes: local command and on line. In the local-command mode, ASCII characters have special meanings. For example, the character "D" followed by a number instructs the modem to dial that number. In the local-command mode, entries from the keyboard travel only as far as the modem; they do not transmit over the phone lines to another computer.

The Smartmodem goes into the on-line mode after answering a call or dialing a call and connecting with a distant modem. In the on-line mode, characters transfer normally through the phone lines to a receiving system. The character "D" no longer causes the modem to dial a number, but transfers as the letter itself.

Eight LEDs on the front panel of the Smartmodem indicate modem-ready, terminal-ready, send-data, receive-data, off-hook, carrier-detect, and auto-answer modes.

Eight DIP switches, accessed behind the front panel of the modem, provide some manual programmability. One switch makes available English responses to command codes. For example, the terminal displays CONNECT after the dialing of a number and the making of a connection with a distant modem. Other settings provide for auto answer on first ring, locally echoed characters and result codes, single-line connection to an RJ11 phone, and an enabling mode for commands.

Micromodem IIE

BAUD: 300
COMPUTER: Apple computers
TYPE: Bell 103
SOFTWARE: Included (software Smartcom I)
AUTO ANSWER: Yes
AUTO DIAL: Yes (pulse- or tone-dial)
PHONE: RJ11 (modular) connection

Features. This internal 300-baud modem inserts into the Apple II, Apple II+, Apple *II*e, or Apple III computers (Fig. 8-13). Working with Smartcom I communications software, it will place and answer calls automatically, create and display files, send and receive files, send data to a printer,

Fig. 8-13. The Micromodem IIE. Courtesy Hayes Microcomputer Products, Inc.

manage communications parameters, and store three telephone numbers for convenient dialing. (See Smartcom I for further information.)

The Micromodem IIE is capable of both pulse- and touch-tone dialing. The small internal speaker allows you to monitor the call, listening for busy signals and wrong numbers.

The modem connects to the phone system through a standard RJ11C connector. A special T-adaptor (not included) can be used to provide an additional connection for a telephone.

Discussion. The Micromodem IIE operates in two functional modes: the terminal mode and the remote-console mode. The terminal mode allows interactive communications between your computer and another computer. The remote-console mode allows your computer to be accessed and controlled from a remote computer or terminal.

The various dialing and configuration commands are executed in the terminal mode, using the control key in conjunction with a character key. On recognizing the control key, the modem processes the additional character entry as a command, rather than passing it through to a receiving computer.

The various configuration and dialing commands can also be executed through Smartcom I. This menu-driven program will prompt for necessary information and automatically send the appropriate control codes to the modem to implement the various modem features.

Through the remote-console mode, commands are entered from the remote system to control an Apple-equipped Micromodem. Information sent to or from either computer displays on the video-display unit of each system; however, the messages from the modem appear only on the local system.

HUMAN ENGINEERED SOFTWARE (HESWARE)

150 North Hill
Brisbane, CA 94005
(415) 468-4110

Hesware has a reputation for quality software products for Commodore 64 and VIC 20 computers. Hesware also directs its hardware line at the Commodore market. The company's hardware products have a 180-day warranty.

The model reviewed and suggested retail price:

Hesmodem I	$69.95

Hesmodem 1

BAUD: 300
COMPUTER: Commodore 64/VIC 20
TYPE: Bell 103
SOFTWARE: Included
AUTO ANSWER: No
AUTO DIAL: No
PHONE: RJ11 (modular) connection

Features. This 300-baud modem is designed for use with either the Commodore 64 or VIC 20. It inserts directly into the user port of the computer and connects to the cable that normally attaches to the handset of a modular phone. The modem includes communications software. It contains a switch for answer and originate mode and a carrier-detect lamp.

The Hesmodem is powered from the computer and requires no additional outlets. After a phone number has been manually dialed, the cable to the handset of a modular phone must be disconnected and then reinserted into the modem for data communications. No additional software is necessary.

The Hesmodem works only with a modular (RJ11C) phone with a full-keystroke number pad.

Discussion. In addition to the modem and software provided, the modem price includes 2 free hours on CompuServe. The procedure for establishing contact with a remote computer includes dialing by hand the number of the service, disconnecting the cable to the handset of the phone, and reconnecting the cable to the modem. This process of dismantling the telephone after each call can be inconvenient. The modem itself hangs precariously from the user port, making the connection vulnerable to abrupt keyboard movements.

The software included with the Hesmodem 1 allows the convenient change of parameters required for RS-232C transmission, including duplex, parity, stop bits, baud rate, and word length. You can select either color or black and white output for your screen.

The software includes a receive-buffer option. At your discretion, you can begin saving information received from the modem. Up to 24K of data can be stored within the receive buffer. That data can be reviewed on the screen or printed on a printer. The modem has no features for saving the buffer to diskette or uploading from diskette.

More sophisticated software, such as Superterm from Midwest Micro, or Vidtex from CompuServe, can be purchased separately and substituted for the program included with the modem. The documentation provides considerable background in telecommunications, as well as needed information for installation and use.

IDEASSOCIATES, INC.

7 Park Dr.
Bedord, MA 01730
(617) 275-4430

IDEAssociates manufactures a number of telecommunications devices, including the IDEAComm 1200. The company markets products directly and through various distributors.

The IDEAComm 1200 is warranted for one year. Simply mail the failed modem card to the manufacturer and IDE Associates will send you an operational board. After the expiration of the warranty period the company will repair the modem on a time and material basis.

Model reviewed and suggested retail price:
IDEAComm 1200 $495.00

IDEAComm 1200

BAUD: 1200/300
COMPUTER: IBM PC/XT
TYPE: Bell 212A/103
SOFTWARE: Included (IDEAComm 1200)
AUTO ANSWER: Yes
AUTO DIAL: Yes (pulse- or tone-dial)
PHONE: RJ11 (modular) connection

Features. The IDEAComm 1200 is an internal modem that inserts into one of the option slots of the IBM PC and XT (Fig. 8-14). It requires that your system have at least one diskette drive, the IBM Personal Computer disk operating system, and disk BASIC language.

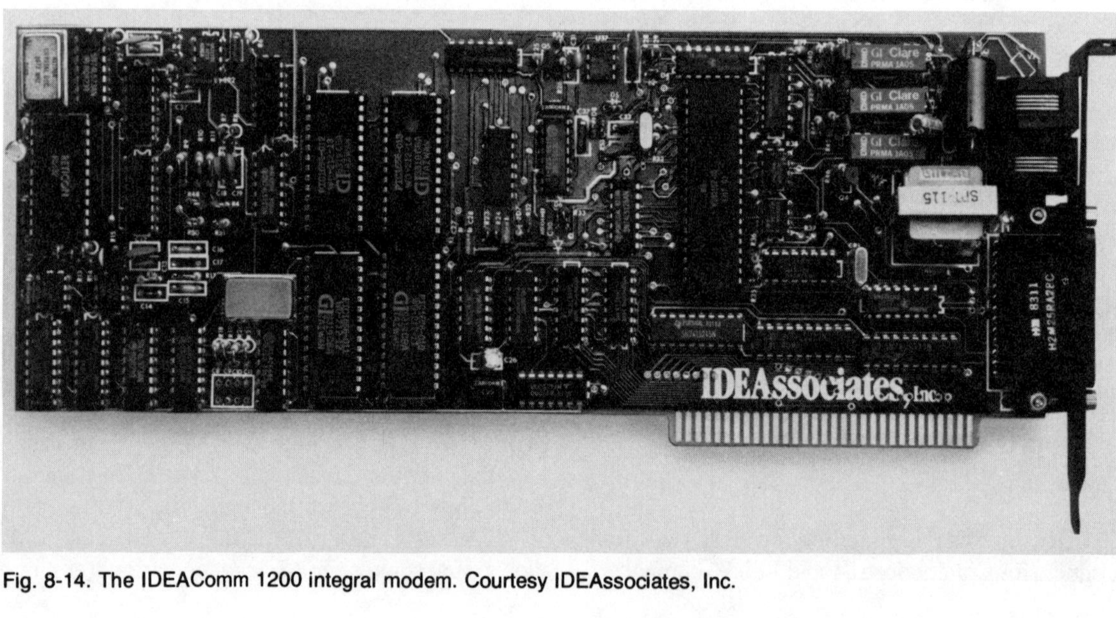

Fig. 8-14. The IDEAComm 1200 integral modem. Courtesy IDEAssociates, Inc.

The modem includes jacks for connecting to an RJ11 phone. One connection is made to the modular wall jack, one to the telephone itself.

The modem includes auto-dial, using either pulse-dial or tone-dial systems. Working with the software, a directory of numbers can be used to automatically dial a number, initiate a logon sequence, and redial if the number is busy.

The modem also includes an auto-answer feature, allowing callers to establish contact and leave messages on an unattended system. The IDEAComm 1200 will automatically display to the caller, in menu format, a list of options for transmitting a message. The caller can look at the disk directory, leave a message, or upload or download a file. An internal speaker allows you to monitor your calls for dial tones, busy signals, etc.

The device includes software providing a menu-driven selection of various modes and features. The device can be used with other communication packages designed to work with the Hayes Smartmodem.

Discussion. The documentation provides thorough instructions on installing the modem into the IBM PC. A jumper setting on the board itself configures the device for either of the two possible serial ports.

Many of the functions of the modem are transparent to the user when using the accompanying software. A menu-driven format provides such options as auto-dial and auto-answer, and will prompt you for necessary data, such as phone numbers or baud rates. Actually, the modem is controlled with single-character commands; most are transmitted by the software automatically in response to your menu selections. A special command mode allows you to transmit some of these control characters directly from the keyboard. These commands are grouped into system commands, dialing commands, and operating commands. A system command "P" followed by a number, for example, will cause the modem to pulse-dial the number. The answering command "A" will cause the modem to answer a telephone immediately, without waiting for a ringing signal. A special control can be implemented to switch a voice call in progress to a call between computers for transmission of data.

A dialing directory of up to 40 names and numbers can be created for executing the auto-dial feature. An up to 30-digit number can be entered. Additional commands necessary for logging on to a system can be included in the directory, as well as specific RS-232C parameters for that number. The logon sequence cannot recognize responses from the other system, but delays can be included for proper timing. Placing a call is as easy as entering the identification number and pressing return. Of course, calls can be made manually from the keyboard without using the directory.

The modem can be configured for RS-232C parameters. A default setting can be established for all transmissions. Individual parameters for each number in the phone directory can be programmed.

Selecting the command mode from the main menu obtains a second menu of options, including the option to have data sent to a printer as it appears on the screen and the option to store all data in a buffer. Periodically, the buffer is transferred to disk.

LEXICON CORPORATION OF MIAMI

1541 N.W. 65th Ave.
Fort Lauderdale, FL 33313
(800) 327-8913

Lexicon, which marketed the first electronic language translator, produces a series of lightweight portable modems. Designed with travel in mind, many include acoustic coupling and optional battery power supply. Lexicon offers to reconfigure its modems for special applications.

The modems are warranted to the original purchaser, for a period of 1 year from the date of purchase, against defects in workmanship or materials under normal use and service. Defective units can be returned to the dealer from whom you made the purchase.

Models reviewed and suggested retail prices:

Lex-11	$145.00
Lex-12	$165.00
Lex-21	$395.00

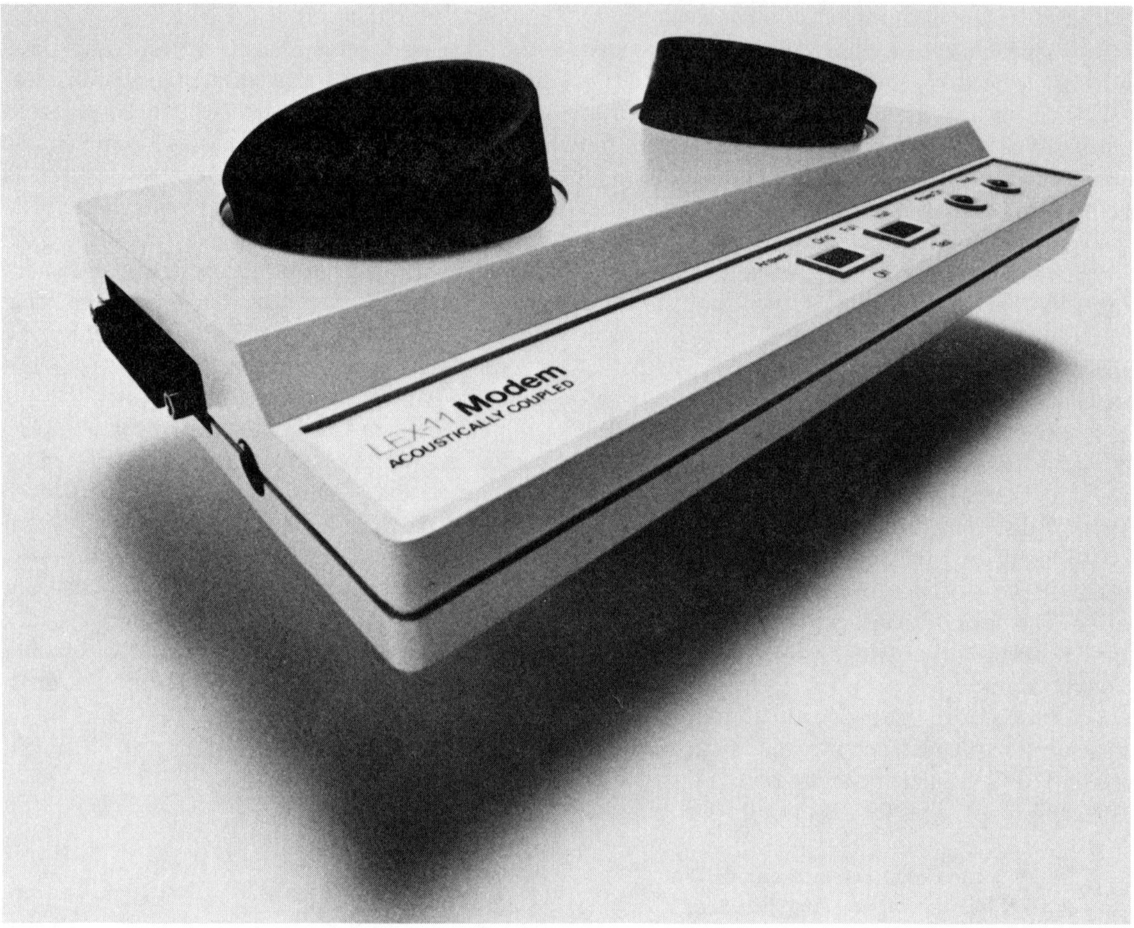

Fig. 8-15. The Lex-11 stand-alone modem. Courtesy Lexicon Corporation of Miami.

Lex-11

BAUD: 300
COMPUTER: Any RS-232C-compatible system
TYPE: Bell 103A
SOFTWARE: Not included
AUTO ANSWER: No
AUTO DIAL: No
PHONE: Acoustic coupled

Features. The Lex-11 is a stand-alone, acoustic-coupled modem operating at speeds of up to 300 bps (Fig. 8-15). Compatible with Bell 103A, it can operate in either full-duplex or half-duplex mode. It connects to a computer through a standard RS-232C interface and is powered through an ac adaptor. An optional battery power supply is available.

A switch setting on the front panel selects from answer mode, originate mode, and off. A second switch selects from half-duplex, full-duplex, and test modes.

Panel LEDs indicate the power-on state and the ready state when the proper carrier signal from a modem at the other end has been detected.

A test mode, by looping transmitted data back to the terminal, verifies that the Lex-11 is functioning properly.

Discussion. This modem is especially well

suited for travel. It is lightweight (3 lbs.) with acoustic coupling. It requires the use of a "G-type" telephone handset, which is pressed firmly into two rubber cups to establish connection through the mouthpiece and earpiece.

The modem can be set for answer mode or originate mode through a switch on the front panel. Likewise, full-duplex or half-duplex modes can be selected from the front panel.

One LED on the front panel indicates that the ac adaptor has been connected to a 120-volt, 60-Hz supply, and that the modem is not in the off position. The other LED illuminates when a call to another compatible modem detects the proper carrier signal.

In the test mode, proper operation of the modem can be verified for both the originate and answer modes. In the answer mode, the modem loops back data, transmitted from your keyboard, to your modem. In the originate mode, proper operation is determined from a tone in the earpiece of the acoustic cushion.

The documentation is rather brief, but sufficient. Of course, operation requires communications software which implements RS-232C protocol for your particular computer.

Lex-12

BAUD: 300
COMPUTER: Any RS-232C-compatible system
TYPE: Bell 103A
SOFTWARE: Not included
AUTO ANSWER: No
AUTO DIAL: No
PHONE: Acoustic coupled or RJ11 (modular connection)

Features. The Lex-12 is a stand-alone, acoustic-coupled modem operating at speeds up to 300 bps. Unlike the Lex-11, it also includes ports for connecting to RJ11 modular-type phones.

It connects to a computer through a standard RS-232C interface and is powered through an ac adapter. An optional battery power supply is available.

A switch on the front panel selects from answer mode, originate mode, and off. A second switch selects from acoustic-coupled, direct-connect (through RJ11), and talk modes. The talk mode allows you to use the telephone for conversation without disconnecting it from the modem. Panel LEDs indicate the power state and the ready state when the proper carrier signal from a modem at the other end has been detected. A test mode, by looping transmitted data back to the terminal, verifies that the Lex-11 is functioning properly.

Discussion. This modem, like the Lex-11, is especially well suited for travel. It is lightweight (3 lbs.) with acoustic coupling. It requires the use of a "G-type" telephone handset, which is pressed firmly into two rubber cups to establish connection through the mouthpiece and earpiece.

The modem can be set for answer mode or originate mode through a switch on the front panel. Likewise, full-duplex or half-duplex modes can be selected from the front panel.

One LED on the front panel indicates that the ac adapter has been connected to a 120-volt, 60-Hz supply, and that the modem is not in the off position. The other LED illuminates when a call to another modem detects the proper carrier signal.

In the test mode, proper operation of the modem can be verified in both the originate and answer modes. In the answer mode, the modem loops back data, transmitted from your keyboard, to your computer. In the originate mode proper operation is determined from a tone in the earpiece in the acoustic cushion.

Operation requires communications software that implements RS-232C protocol for your particular computer.

Lex-21

BAUD: 300
COMPUTER: Self-contained terminal printer
TYPE: Bell 103A
SOFTWARE: Included
AUTO ANSWER: No

Fig. 8-16. The Lex-21 portable terminal and printer has a built-in modem. Courtesy Lexicon Corporation of Miami.

AUTO DIAL: No
PHONE: RJ11 (modular connection)

Features. The Lex-21 is a lightweight (5 lbs.), portable terminal and printer with a built-in, 300-baud modem (Fig. 8-16). The modem is connected through a standard RJ11 (modular) connection, but an optional acoustic coupler is available.

The keyboard has 59 keys. An optional numeric keypad is also available. Commands for configuring the modem, as well as text, can be entered through the keyboard.

The Lex-21 has a 1K receive buffer and a 2K internal memory for composing and editing text. Text can be entered through a keyboard and stored in nonvolatile memory. Turning the device off or using it to transmit data over the phone lines does not affect the stored text. The text can also be edited, using commands similar to most word processors. The text can be printed on the 40-column thermal printer or transmitted through the modem.

Discussion. The Lex-21 is similar in appearance to a typewriter. In the editing mode, it behaves like a typewriter—characters entered from the keyboard print on the printer. Data entered from the keyboard is stored within a 2K internal memory. Text stored in the memory can be printed at any time and modified with special function keys. A pointer can be moved to the beginning or ending of text and then moved one line at a time. Characters, words, and lines following the pointer can be erased from memory or printed.

A second mode of operation allows the configuration of RS-232C parameters through keyboard entries. In this mode, the various parameter settings are listed on the printer. The various settings, including baud rate, parity, duplex, and stop bits, can be changed by pressing appropriate keys on the keyboard. These settings remain even after the terminal is turned off.

A third mode of operation, the on-line mode, allows data communications with another compatible device through the phone lines. Data received in this mode is printed on the printer. Data can be transmitted from the keyboard and printed as well. Any text previously saved in memory through the editor mode can be transmitted in the on-line mode.

Switches on the Lex-21 select between answer or originate mode, and between acoustic and direct connections. A series of LEDs indicates the power status, the on-line status, the memory status, the carrier-detect signal, and whether data is being received.

MFJ ENTERPRISES, INC.

P.O. Box 494
Mississippi State, MS 39762
1-800-647-1800

MFJ Enterprises manufactures and sells amateur radio accessories. Additionally, the company markets some peripherals for the Commodore 64 and VIC 20 computers. Marketing mainly through mail order, MFJ accepts, through the toll-free number just listed, orders for products and requests for product catalogs.

The company offers an unconditional warranty on most of its products for 1 year from date of purchase. It will repair or replace any product that fails or is defective in the first year. If for any reason you are dissatisfied with any product, you can return it during the first 30 days for a full refund.

Models reviewed and suggested retail prices:

1237 Modem	$ 49.95
1233 Modem	$129.95

1237 Modem

BAUD: 300
COMPUTER: Commodore 64, VIC 20
TYPE: Bell 103
SOFTWARE: Includes printed listing for terminal software
AUTO ANSWER: No
AUTO DIAL: No
PHONE: RJ11 (modular) connection

Features. This Bell 103-compatible modem is designed for use with either the Commodore 64 or VIC 20. It inserts directly into the user port of the computer and must be connected to the cable that normally connects to the handset of a modular phone. A BASIC listing of a terminal program is included with the documentation. The program can be entered through the keyboard and saved for later use. A terminal program can be purchased separately—on tape for $4.95, or on cartridge for $19.95. The modem contains a single switch for answer and originate modes. It also has a carrier-detect lamp.

The modem has the essentials necessary to begin telecomputing. It is powered from the computer and requires no additional outlets. The cable to the handset of a modular phone must be disconnected and then reinserted into the modem for data communications.

The modem works only with a modular (RJ11C) phone with a full-stroke keypad.

Discussion. The modem provides the essentials for telecomputing at a very affordable price. This product has, however, some distinct limitations. For example, to establish contact with a remote computer, you must dial by hand the number of the service, disconnect the cable to the handset of the phone, and reconnect the cable to the modem. This process of dismantling the telephone each time can be inconvenient. The modem itself hangs precariously from the user port and presents the risk of disconnection.

1233 Modem

BAUD: 300
COMPUTER: Any RS-232C-compatible system
TYPE: Bell 103
SOFTWARE: Not included
AUTO ANSWER: No

AUTO DIAL: No
PHONE: RJ11C (modular) connection

Features. This modem provides both direct-connect and acoustic-coupled connections to the telephone system. One switch on the modem toggles between full-duplex mode and half-duplex mode. A second switch allows the user to alternate between originate and answer modes. A button on the front panel performs a self-test of the modem.

The modem provides connections for standard RS-232C devices. One port connects directly to a DB25 connector; a second port provides an interface to TTL interfaces, such as the Commodore 64 user port.

The modem is powered from a 12-volt dc adapter that connects to standard 115-volt ac wall sockets. The modem may also be powered through two 9-volt batteries.

Discussion. Using a direct connection, the modem provides one outlet for connection to the handset of a modular phone. A second outlet provides a connection with the base of the telephone. If the phone is not a modular type, the handset can be inserted into the two cups for an acoustic-coupled connection.

In addition to the standard DB25 connection, a five-pin socket allows devices with TTL voltages to be connected directly to the modem. The five lines are: data out, data in, ground, carrier-detect, and an unused pin.

A button on the front panel performs a self-test of the internal circuitry of the modem. With test button in, characters can be entered through the keyboard of the computer or terminal. The characters are displayed on the screen, verifying the working condition of the modem and connecting circuitry.

A power LED on the front panel indicates proper supply voltages, and a carrier-detect LED indicates a valid carrier signal from a remote modem.

MICROBITS PERIPHERAL PRODUCTS, INC.

225 W. Third St.
Albany, OR 97321
(503) 967-9075

Microbits Peripheral Products warrants the MPP-1000C Modem and included cartridge to be free from defects in materials and workmanship and to meet applicable specifications under normal use and service for 1 year after sale to the original owner.

Model reviewed and suggested retail price:

MPP-1000C $149.95

MPP-1000C

BAUD: 300
COMPUTER: ATARI
TYPE: Bell 103
SOFTWARE: Included
AUTO ANSWER: Yes
AUTO DIAL: Yes
PHONE: RJ11 (modular) connection

Features. This Bell 103-compatible modem is designed for use with any ATARI computer (Fig. 8-17). It connects directly to the joystick port and requires no additional interface. The modem includes software and supports disk drives, cassette recorders, and printers.

It is capable of transmitting at rates of up to 300 baud and has auto-dial, auto-answer, originate, and answer modes. The modem connects directly to a modular phone set (RJ11). Use of the phone for voice or data communications can be conveniently changed with a single switch setting on the modem.

The software supports XMODEM protocol and automatically makes the translation between ASCII and ATASCII characters. Text can be uploaded, downloaded, or streamed directly to a printer, avoiding the constraints of a limited memory.

Discussion. The software "Smart Terminal," included with the modem, inserts into the cartridge slot of the computer. It is menu driven in format and allows configuration of the various modes of operation of the modem.

A buffer mode allows the creation of up to nine separate buffers for the temporary storage of data. The menu displays the remaining space in memory for additional data. Information can also be downloaded from the modem directly onto diskette,

Fig. 8-17. The MPP-1000C modem. Courtesy Microbits Peripheral Products, Inc.

to avoid using buffer space.

The print feature allows you to access information from a host computer and send the information directly to a printer. Because the baud rate of the transmission is 300, the printer must be capable of printing about 30 characters per second; so the feature is unsuitable for many daisy-wheel printers. With a properly interfaced printer, however, the necessity for downloading, with its limitations, is avoided. A single switch implements the printing feature.

The auto-answer feature of the modem allows you to leave the computer unattended and receive messages from other computers automatically. This mode can also automatically upload a file and hang up the phone when completed.

The auto-dial feature allows you to enter a phone number from the keyboard of your computer and have that number dialed automatically. You can also select a phone number from a list of 10 numbers for automatic dialing.

The XMODEM protocol, used by many bulletin boards, can be implemented to automatically correct any transmission errors. Additional options include 38- or 40-column displays, depending on your type of video-display unit. You can have an 80-column display if your system has an 80-column board.

MORROW, INC.

600 McCormick St.
San Leandro, CA 94577
(415) 430-1970

Morrow makes and markets a comprehensive product line of business systems and computers.

The company has over 600 authorized dealers nationwide. Hardware is warranted for 90 days from the date of purchase. In addition to the 600 authorized dealers, service can be acquired at any of the over 80 Xerox third-party service centers.

The model reviewed and suggested retail price:

MM300 $299.00

MM300

BAUD: 300
COMPUTER: Micro Decision Computers
TYPE: Bell 103/113
SOFTWARE: Included (ReachOut)
AUTO ANSWER: Yes
AUTO DIAL: Yes (pulse- or tone-dial)
PHONE: RJ11 (modular) connection

Features. This Bell 103/113-compatible modem is designed for use with the Morrow Micro Decision Computer (Fig. 8-18). It is capable of transmitting at rates of up to 300 baud and has auto-dial, auto-answer, originate, and answer modes. The modem connects to the rear of the computer and includes a diskette for the communi-

Fig. 8-18. The MM300 modem is designed for use with the Morrow Micro Decision Computer. Courtesy Morrow, Inc.

cations software *ReachOut*. All of the features of the modem operate through the menu-driven format of the software. It is difficult to discuss the modem without discussing the software.

Dialing can be performed manually or by selecting from a list of choices in a telephone-directory file. The directory can be set up to redial the number, if necessary, and to wait a specified amount of time before giving up. Special "script" files can be included in the directory to execute a logon sequence.

The software includes features for transferring files individually or as a batch. Between two systems using ReachOut software, an error-detection feature will automatically retransmit data until it is received correctly.

Discussion. The software displays four menus: a dialing menu controls the dialing functions; the control-mode menu controls the major modes of operation, the configuration menu permits you to change the operating characteristics of the modem, and the protocol menu allows you to change from one type of error-checking protocol to another.

A dialing directory can contain up to 26 numbers and is created with any word-processor program. Logon sequences can be created as special sections in the dialing directory called "scripts." Scripts can be up to 80 characters long and include delays for dial tones, etc. Each of the characters in a script executes a particular command. Included in the commands is a "wait for" character that will monitor the data received from another modem and act only when it receives a particular number or character. This feature is very useful in a logon sequence where you must respond to various questions.

A logging-on mode allows you to download and save information received from a distant modem. The logging-on mode can be disabled intermittently to save selected material.

Files can be uploaded individually or in batches. A batch file, which lists numerous other files, can be made with a word-processing program. When the batch file is executed, ReachOut will automatically upload each file, one at a time.

The various RS-232C parameters, such as baud and parity, as well as the modes of operation, are implemented through software control.

A protocol menu allows selection of ReachOut-to-ReachOut protocol, "Christensen" protocol, CP /M User's Group, and Echo-plex. Each selection has its own error-detection scheme.

ReachOut includes two public-domain programs that compress binary files by 10 percent and text files by 40 percent. Once transmitted, the files can be expanded to their original form. This process reduces the time required for transmission.

All features of the Morrow Micro Decision computer can be used normally. The information displayed on the screen can be sent to a printer or stored on diskette.

MULTI TECH SYSTEMS, INC.

82 Second Ave. S.E.
New Brighten, MN 55112
(612) 631-3550

Multi Tech Systems manufactures communications equipment, including 300-, 1200-, and 2400-baud modems, acoustic data couplers, interface testing devices, statistical multiplexers, and data-access arrangements. They are sold through a nationwide network of distributors, representatives, dealers, and retailers.

Multi Tech warrants, to the original user, that its products are free from defects in workmanship and material for a period of 1 year (2 years for some MultiModems) from the original date of shipment from the factory in New Brighten. After the warranty expires, service can be obtained through the factory with a standard repair charge of about $70.

Models reviewed and suggested retail prices:

MultiModem PC	$549.00
MultiModem HC3	$289.00
MultiModem	$549.00
MultiModem HC	$549.00

MultiModem PC

BAUD: 1200/300

Fig. 8-19. The MultiModem PC. Courtesy Morrow, Inc.

COMPUTER: IBM PC, IBM XT, IBM PC-compatible
TYPE: Bell 212A/103/113
SOFTWARE: Included
AUTO ANSWER: Yes
AUTO DIAL: Yes (pulse- or tone-dial)
PHONE: RJ11 (modular) connection

Features. The MultiModem PC is an internal modem that inserts into the full-size slot in the IBM PC or IBM XT (Fig. 8-19). The modem operates at 1200 baud in a Bell 212A format or at 0-300 baud in a Bell 103 format. The modem is also compatible with Racal-Vadic "triple" modems, which are Bell 212A compatible.

Commands are Hayes compatible and provide for touch-tone dialing, pulse-dialing, and repeat dialing. The modem also includes an auto-answer feature which allows callers to establish contact and leave messages on an unattended system. The modem includes circuitry for analog loopback and self-tests.

Discussion. The various features of the modem are conveniently accessed through the software included with the modem: Multi-Com. It includes a terminal mode with full cursor control and terminal emulation. Phone numbers can be dialed directly or from a menu of up to 20 numbers.

Data communication may be captured into a designated file. Text files can be listed to a CRT or printer. Any text or binary file may be received or transmitted using a protected-protocol transfer.

Automatic logon macro commands may be specified by the user to include phone number, timeouts, search for strings, output strings, pauses, macro-authorization password, and display protection to prevent unauthorized access. Communication parameters and logon macros may be specified individually for each of the 20 phone numbers. System parameters and user menus are stored in a parameter file. This file may be edited to customize the program for special applications. Translation tables allow automatic conversion of any key on the keyboard to any single character or string of characters for transmission. Similarly, incoming characters can be translated for display on the screen.

The MultiModem PC is designed to recognize

one- and two-character commands from the keyboard. These commands configure the modem and execute dialing and answering features.

MultiModem HC3

BAUD: 300
COMPUTER: Any RS-232C-compatible system
TYPE: Bell 103
SOFTWARE: Not included
AUTO ANSWER: Yes
AUTO DIAL: Yes (pulse- or tone-dial)
PHONE: RJ11 (modular) connection

Features. The MultiModem HC3 is an auto-originate, answer or auto-answer, half-duplex or full-duplex modem (Fig. 8-20). It operates asynchronously at 300 baud in a Bell 103 format. The MultiModem HC3 provides automatic dialing and automatic answering. The commands that control the modem are fully compatible with the Hayes command set; all communications software compatible with the Hayes Smartmodem 1200 should work with the MultiModem HC3.

The MultiModem connects to most RS-232C-compatible systems through a standard DB25 female connector. Two RJ11C connections allow connection to the nearest phone outlet and to a compatible telephone.

In addition to the auto-dial feature, the last number dialed can be redialed automatically. An answering mode programs the modem to answer an incoming call automatically, without attention. The modem can be programmed to answer either immediately or after a certain number of rings.

Several switches, accessed by removing the modem's front cover, allow hardware configuration. Diagnostic capabilities include a local analog loop-back test and seven LED status indicators.

Discussion. The MultiModem operates in one of two functional states: the command mode and the on-line mode. The modem assumes the command mode on being first powered up. In the command mode, the modem recognizes one- or two-character commands entered from the computer keyboard and implements the appropriate features.

When it has dialed a number and established contact with a remote computer, the modem switches to the on-line mode. In this mode, characters which were commands in the command mode

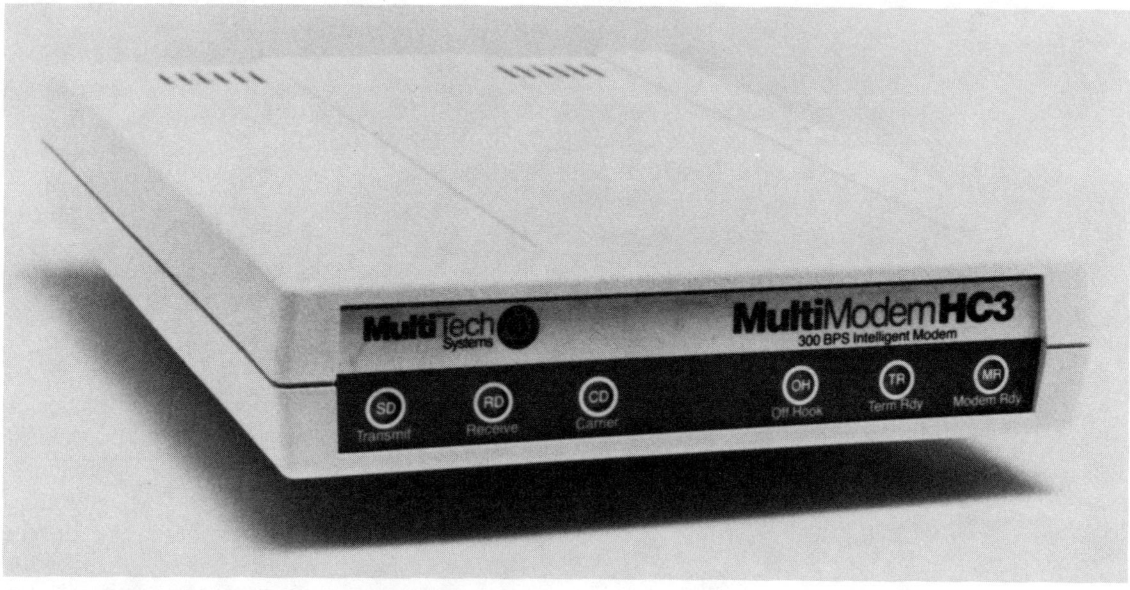

Fig. 8-20. The MultiModem HC3. Courtesy Morrow, Inc.

pass on normally to the receiving system. On disconnecting from a remote system, the modem automatically resumes the command mode.

A series of internal switches enable various configuration options. They include: a forced DTR, for terminals that do not provide a DTR signal; verbal responses, providing messages on the status of the modem (e.g., RINGING, BUSY, etc.); a command echo, providing a display of the commands entered from the keyboard; auto-answer; forced carrier-detect and DSR signals, for terminals that require these signals at all times; RJ11 option, ensuring compatibility with RJ11, RJ12, or RJ13 phone lines; and command-mode disable, for applications which need to have the command set ignored at all times.

A series of LEDs provide information on the operation and status of the modem. They include: transmit data, indicating that data is being transmitted; receive data, indicating that data is being received; carrier detect, indicating that a valid carrier signal has been detected; off hook, indicating that the modem is dialing; on line or answering a call; terminal ready, indicating that the terminal is ready; and modem ready, indicating the modem has power and is ready.

MultiModem

BAUD: 1200/300
COMPUTER: Any RS-232C-compatible system
TYPE: Bell 212/103
SOFTWARE: Not included
AUTO ANSWER: Yes
AUTO DIAL: Yes (pulse- or tone-dial)
PHONE: RJ11 (modular) connection

Features. The MultiModem is an auto-originate, answer or auto-answer, half-duplex or full-duplex modem (Fig. 8-21). It operates asynchronously at 300 baud in a Bell 103 format and at 1200 baud in a Bell 212A format. The MultiModem provides automatic dialing and is compatible with nearly all communications software which uses Hayes commands. Two exceptions are Hayes Smartcom II, for the IBM PC, and Epson's Valdocs, for the QX10.

Up to 6 telephone numbers, of up to 31 digits each, can be stored in the modem's battery-backed memory. The modem can either pulse-dial or tone-dial.

The MultiModem connects to most RS-232C-

Fig. 8-21. The MultiModem. Courtesy Morrow, Inc.

compatible systems through a standard DB25 female connector. Two RJ11C connections allow hook up to the nearest phone outlet and to a compatible telephone.

In addition to the auto-dial feature, numbers can be redialed automatically. If the number is busy, a linking feature can automatically dial a second number. The MultiModem can recognize a busy signal and immediately hang up or redial the same or another number.

An answering mode programs the modem to answer an incoming call automatically, without attention. The modem can be programmed to answer either immediately or after a certain number of rings.

The modem is designed to be compatible with software implementing Hayes commands. Additional features, not included in the Hayes command set, are implemented with commands that do not conflict with Hayes.

Several switches, accessed by removing the modem's front cover, allow hardware configuration. Diagnostic capabilities include a local analog loopback test and seven LED status indicators.

Discussion. The MultiModem operates in two functional states: the command mode and the on-line mode. The modem assumes the command mode when it is first powered up. In the command mode, the modem recognizes one- or two-character commands entered from the keyboard and implements the appropriate features.

When it has dialed a number and established contact with a remote computer, the modem switches to the on-line mode. In this mode, characters which were commands in the command mode pass on normally to the receiving system. On disconnecting from a remote system, the modem automatically resumes the command mode.

A series of commands allow the user to enter up to 6 numbers into the modem's memory. The number can be automatically dialed by entering the characters "ATNN," where n is a number from 1 to 6 representing the particular phone number desired. The modem can be programmed to redial continuously if a number is busy. If desired, a secondary number can be dialed. Normally, the modem waits for a dial tone before dialing, but a blind-dialing feature enables it to ignore the dial tone for a system that has a nonstandard tone.

A series of internal switches enable various configuration options. They include: forced DTR, for terminals that do not provide a DTR signal; verbal responses that provide English messages on the status of the modem (e.g., RINGING, BUSY, etc.); command echo, providing a display of the commands entered from the keyboard; auto-answer; forced carrier-detect and DSR signals, for terminals that require these signals at all times; RJ11 option, ensuring compatibility with RJ11, RJ12, or RJ13 phone lines; and command-mode disable, for applications which need to have the command set ignored at all times.

A series of LEDs provides information on the operation and status of the modem. They include: transmit data, indicating data is being transmitted; receive data, indicating that data is being received; carrier detect, indicating that a valid carrier signal has been detected; high speed, indicating 1200-baud operation; off hook, indicating the modem is dialing; on line or answering a call; terminal ready, indicating that the terminal is ready; and modem ready, indicating that the modem has power and is ready.

MultiModem HC

BAUD: 1200/300
COMPUTER: Any RS-232C-compatible system
TYPE: Bell 212/103
SOFTWARE: Not included
AUTO ANSWER: Yes
AUTO DIAL: Yes (pulse- or tone-dial)
PHONE: RJ11 (modular) connection

Features. The MultiModem HC is an auto-originate, answer or auto-answer, half-duplex or full-duplex modem (Fig. 8-22). It operates asynchronously at 300 baud in a Bell 103 format, and at 1200 baud in a Bell 212A format. The MultiModem provides automatic dialing and automatic answering. The commands that control the modem are fully compatible with the Hayes command set; all com-

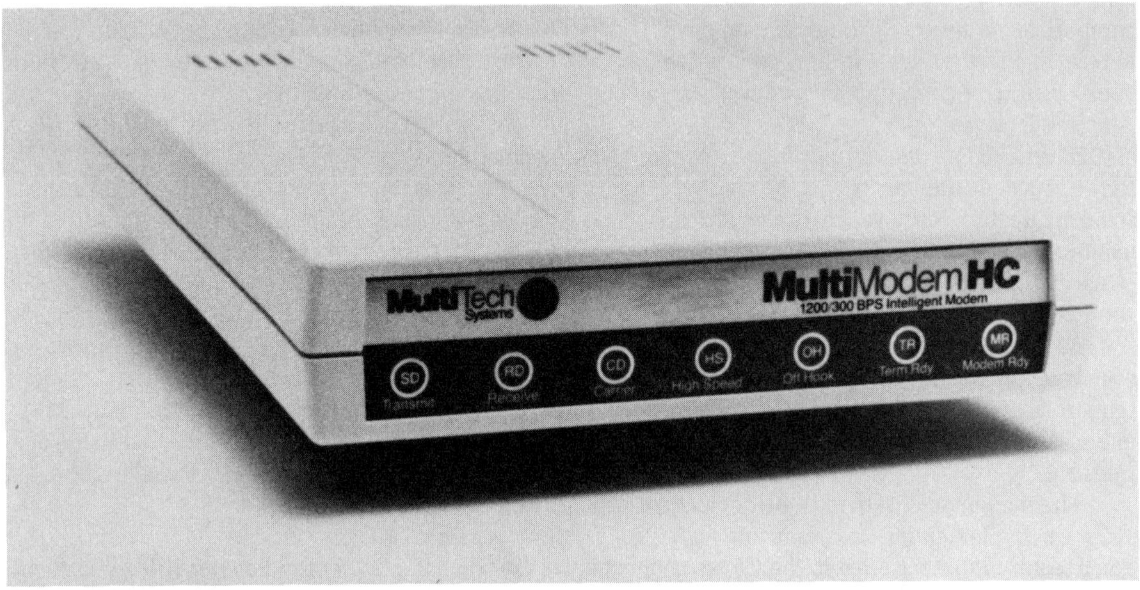

Fig. 8-22. The MultiModem HC. Courtesy Morrow, Inc.

munications software that is compatible with the Hayes Smartmodem 1200 will work with the MultiModem HC.

The MultiModem connects to most RS-232C-compatible systems through a standard DB25 female connector. Two RJ11C connections allow hookup to the nearest phone outlet and to a compatible telephone.

In addition to the auto-dial feature, the last number dialed can be redialed automatically. An answering mode programs the modem to answer an incoming call automatically, without attention. The modem can be programmed to answer either immediately or after a certain number of rings.

Several switches, accessed by removing the modem's front cover, allow hardware configuration. Diagnostic capabilities include a local analog loop-back test and seven LED status indicators.

Discussion. The MultiModem HC operates in two functional states: the command mode and the on-line mode. The modem assumes the command mode when it is first powered up. In the command mode, the modem recognizes one- or two-character commands entered from the keyboard, and implements the appropriate features.

When it has dialed a phone number and established contact with a remote computer, the modem switches to the on-line mode. In this mode, characters that were commands in the command mode pass on normally to the receiving system. On disconnecting from the remote system, the modem automatically resumes the command mode.

A series of internal switches enable various configuration options. They include: a forced DTR for terminals that do not provide a DTR signal; verbal responses, providing English messages on the status of the modem (e.g., RINGING, BUSY, etc.); command echo providing a display of the commands entered from the keyboard; auto-answer; forced carrier-detect and DSR signals, for terminals that require these signals at all times; RJ11 option, ensuring compatibility with RJ11, RJ12, or RJ13 phone lines; and command-mode disable, for applications which need to have the command set ignored at all times.

A series of LEDs provide information on the operation and status of the modem. They include: transmit data, indicating data is being transmitted; receive data, indicating data is being received; carrier detect, indicating that a valid carrier signal has been detected; high speed, indicating 1200-baud operation; off hook, indicating the modem is dialing;

on line or answering a call; terminal ready, indicating that the terminal is ready; and modem ready, indicating that the modem has power and is ready.

NEC AMERICA, INC.

1012 Stewart Dr.
Sunnyvale, CA 94086
(408) 737-7711

NEC produces a number of data communications products, including high speed modems for private line applications. The company's modems are warranted against defects in parts and workmanship for 2 years.

Model reviewed and suggested retail price:

N212BR $795.00

N212BR

BAUD: 1200/300
COMPUTER: Any RS-232C-compatible system
TYPE: Bell 212A
SOFTWARE: Not included
AUTO ANSWER: Yes
AUTO DIAL: Yes (tone- or pulse-dial)
PHONE: RJ11 (modular) connection

Features. The N212BR is a Bell 212A-compatible modem which is plug compatible with the Bell 212AR stand-alone and multiple-mounting configurations. Telephone numbers can be stored in any of 12 directory locations together with associated logon messages (up to 64 characters per location). Numbers are stored in nonvolatile, battery-protected memory within the modem.

Telephone numbers can be dialed from directory entries, from a specified list of directory numbers, or from the terminal keyboard. The last number can be automatically redialed.

Logon messages stored in the directory can be protected so that they will not print, either during transmission or when displaying memory. Diagnostic capabilities include an analog loopback, digital loopback, self-test, and seven LED status indicators.

Discussion. One- and two-character commands, entered from the keyboard of a terminal or computer, execute the various features and modes of the modem. Editing commands enable phone numbers, logon messages, and other commands to be entered into the modem's memory.

Numbers in the directory can be dialed using the one- or two-digit directory-location numbers. Telephone numbers in memory can also be dialed in a specified sequence.

The user may program a password into the modem to prevent unauthorized use of the modem or unauthorized access to telephone numbers and logon messages.

A series of LEDs provide information on the operation and status of the modem. They include: send data, receive data, terminal ready, modem ready, make busy, high speed, modem check, and test mode.

NOVATION INC.

18664 Oxnard St.
Tarzana, CA 91356
(213) 996-5060

Novation warrants its products to be free from defects in parts and workmanship for a period of 1 year. The company has a free 24-hour, 300-baud computer that provides modem and printer tests. Technical questions can be submitted through electronic mail.

Modems reviewed and suggested retail prices:

103 Smart-Cat	$249.00
103/212 Smart-Cat	$595.00
D Cat	$199.00
J Cat	$149.00
Cat	$189.00
212 Auto-Cat	$695.00
Apple Cat II	$595.00 with upgrade

103 Smart-Cat

BAUD: 300

Fig. 8-23. The 103 and 103/212 Smart-Cat modems. Courtesy Novation, Inc.

COMPUTER: Any RS-232C-compatible system
TYPE: Bell 103
SOFTWARE: Not included
AUTO ANSWER: Yes
AUTO DIAL: Yes (pulse- or tone-dial)
PHONE: RJ11 (modular) connection

Features. The Smart-Cat is an external modem that connects through a standard DB-25 connector with RS-232C-compatible systems (Fig. 8-23). The modem operates at 0 to 300 baud in a Bell 103-compatible format.

Commands provide for touch-tone dialing, pulse-tone dialing, and repeating dialing instructions. The modem also includes an auto-answer feature which allows callers to establish contact and leave messages on an unattended system. Features and configurations of the Smart-Cat are implemented through commands entered from the keyboard of the computer or terminal. The modem includes circuitry for both analog and digital loopback tests.

Discussion. A set of five switches, accessed by removing the front cover of the modem, determine the default state of certain commands. One switch assigns the special character that sets the modem into the command mode; the character is either "%" or CONTROL-N. A second switch provides long or short messages in response to commands entered through the keyboard. In short form, numbers replace words like BUSY, OK, READY, etc. A third switch determines the data rate. A fourth switch forces the DSR line high all the time. A fifth switch provides a dynamic carrier detect required by some communications programs.

Strings of one- and two-character commands are entered from the keyboard to implement various modes and functions of the modem. A single programmed character (any ASCII value) can be used to switch from the command mode, where keyboard entries are interpreted as commands, to the on-line mode, where data is transmitted to a receiving modem.

A series of tests implemented through software control performs analog and local digital loopback tests.

103 / 212 Smart-Cat

BAUD: 1200/300
COMPUTER: Any RS-232C-compatible system

TYPE: Bell 212A/103/113
SOFTWARE: Not included
AUTO ANSWER: Yes
AUTO DIAL: Yes (pulse- dial/tone-dial)
PHONE: RJ11 (modular) connection

Features. This Smart-Cat is an external modem that connects through a standard DB-25 connector with RS-232C-compatible systems (Fig. 8-23). The modem operates at 1200 baud in a Bell 212A-compatible format or at 0 to 300 baud in a Bell 103-compatible format.

Commands provide for touch-tone dialing, pulse-tone dialing, and repeating dialing instructions. The modem also includes an auto-answer feature, allowing callers to establish contact and leave messages on an unattended system. Features and configurations of the Smart-Cat are implemented through commands entered from the keyboard of the computer or terminal. The modem includes circuitry for both analog and digital loopback tests.

Discussion. A set of five switches, accessed by removing the front cover of the modem, determines the default state of certain commands. One switch assigns the special character that sets the modem into the command mode; the character is either "%" or CONTROL-N. A second switch provides long or short messages in response to commands entered through the keyboard. In the short form, numbers replace words like BUSY, OK, READY, etc. A third switch determines the data rate as either high speed (1200 baud) or low speed (0-300 baud). A fourth switch forces the DSR line high all the time. A fifth switch provides a dynamic carrier detect required by some communications programs.

Strings of one- and two-character commands are entered from the keyboard to implement various modes and functions of the modem. A single programmed character (any ASCII value) can be used to switch from the command mode, where keyboard entries are interpreted as commands, to the on-line mode, where data is transmitted to a receiving modem. A series of tests implemented through software control performs analog, remote digital, and local digital loopback tests.

D Cat Modem

BAUD: 300
COMPUTER: Any RS-232C-compatible system
TYPE: Bell 103/113
SOFTWARE: Not included
AUTO ANSWER: No
AUTO DIAL: No
PHONE: RJ11 (modular) connection

Features. The D Cat is an external 300-baud modem that connects through a standard DB-25 connector to RS-232C-compatible system (Fig. 8-24). The handset of a modular phone is connected to the modem, and a second cable connects the modem to the phone. Power is supplied to the D-Cat through an ac adapter that plugs into a standard 115-volt ac outlet.

Switch settings on the modem configure it for answer mode or originate mode, full duplex or half duplex, and talk mode or data mode. In data mode, the telephone is used to transmit data; in talk mode, the phone can be used for normal conversation.

The modem includes an analog-loopback test to verify the working condition of the system. LEDs indicate that the power is on, and the modem is ready to transmit data.

Discussion. Communication through the D-Cat is initiated in the talk mode by manually dialing a connected phone and waiting for the answering tone transmitted by the receiving system. The modem is then switched to data mode, and transmission of data can begin.

An analog self-test loops data transmitted through the modem back to the screen to verify that the modem is working properly. Additional communications software for your particular system is necessary.

J-Cat Modem

BAUD: 300

Fig. 8-24. The D Cat modem. Courtesy Novation, Inc.

COMPUTER: Any RS-232C-compatible system
TYPE: Bell 103/113
SOFTWARE: Not included
AUTO ANSWER: Yes
AUTO DIAL: Yes (pulse-dial)
PHONE: RJ11 (modular) connection

Features. The J-Cat is an external 300-baud modem that connects through a standard DB-25 connector to RS-232C-compatible systems (Fig. 8-25). The modem has a single RJ11C socket for direct connection to an appropriate wall outlet. The connection can include a modular phone using a Y adapter. Power is applied to the modem through a standard ac adapter included with the modem. This modem automatically configures itself to originate or answer mode by detecting the carrier signal from the other modem. A panel switch on the modem enables a keyboard connect feature. Pressing any key on the keyboard will then set the modem to an off-hook condition.

Another switch will automatically transmit a "break" to the remote computer which will interrupt the data transmission without losing the carrier. The J-Cat can be set to automatically answer an incoming call.

The modem includes an analog loopback test for verification of the working condition of the system. LEDs indicate that the power is on, and the modem is in the off-hook condition.

Discussion. Communication through the J-Cat is initiated manually by dialing the number through the telephone and then putting the modem in the off-hook condition. To use the auto-dial feature of the modem, some special programming described in the documentation is necessary.

An analog self-test loops data transmitted through the modem back to the screen to verify that the modem is working properly. Additional communications software for your particular system is necessary.

Cat Modem

BAUD: 300

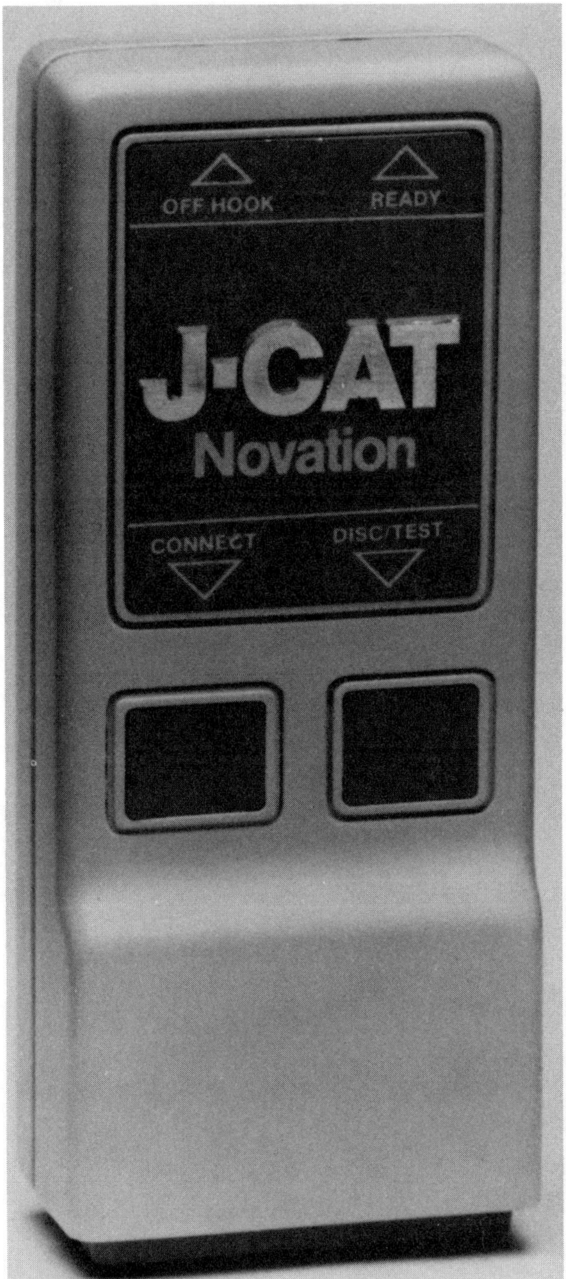

Fig. 8-25. The J-Cat modem. Courtesy Novation, Inc.

COMPUTER: Any RS-232C-compatible system
TYPE: Bell 103/113
SOFTWARE: Not included
AUTO ANSWER: No
AUTO DIAL: No
PHONE: Acoustic coupled connection

Features. The Cat is an external 300-baud modem that connects through a standard DB-25 connector to RS-232C-compatible systems (Fig. 8-26). The modem is acoustically coupled to a standard telephone. Power is supplied to the Cat through an ac adapter that plugs into a standard 115-volt ac outlet.

Switch settings on the modem configure it for answer mode or originate mode, and for full-duplex or half-duplex. The modem includes an analog loop-back test for verification of the working condition of the system. LEDs indicate that the power is on, and the modem is ready to transmit data.

Discussion. Communication through the Cat is initiated in the talk mode by manually dialing a phone and waiting for the answering tone transmitted by the receiving system. The handset of the phone is then inserted into acoustic muffs on the modem itself. When the READY LED turns on, the modem is ready to communicate.

An analog self-test loops data transmitted through the modem back to the screen to verify that the modem is working properly. Additional communications software for your particular system is necessary.

212 Auto Cat

BAUD: 1200/300
COMPUTER: Any RS-232C-compatible system
TYPE: Bell 212A/103/113
SOFTWARE: Not included
AUTO ANSWER: Yes
AUTO DIAL: Yes (pulse-dial)
PHONE: RJ11 (modular) connection

Features. The Auto Cat is an external modem that connects through a standard DB-25 connector to RS-232C-compatible systems (Fig. 8-27). The modem operates at 1200 baud in a Bell 212A format or at 0 to 300 baud in a Bell 103 format.

Fig. 8-26. The Cat modem. Courtesy Novation Inc.

The modem includes circuitry for both analog and digital loopback tests. Additional features include auto-answer and auto-dial (requires special programming).

Discussion. Ten option switches configure the modem for various applications. They include:

Fig. 8-27. The 212 Auto Cat. Courtesy Novation Inc.

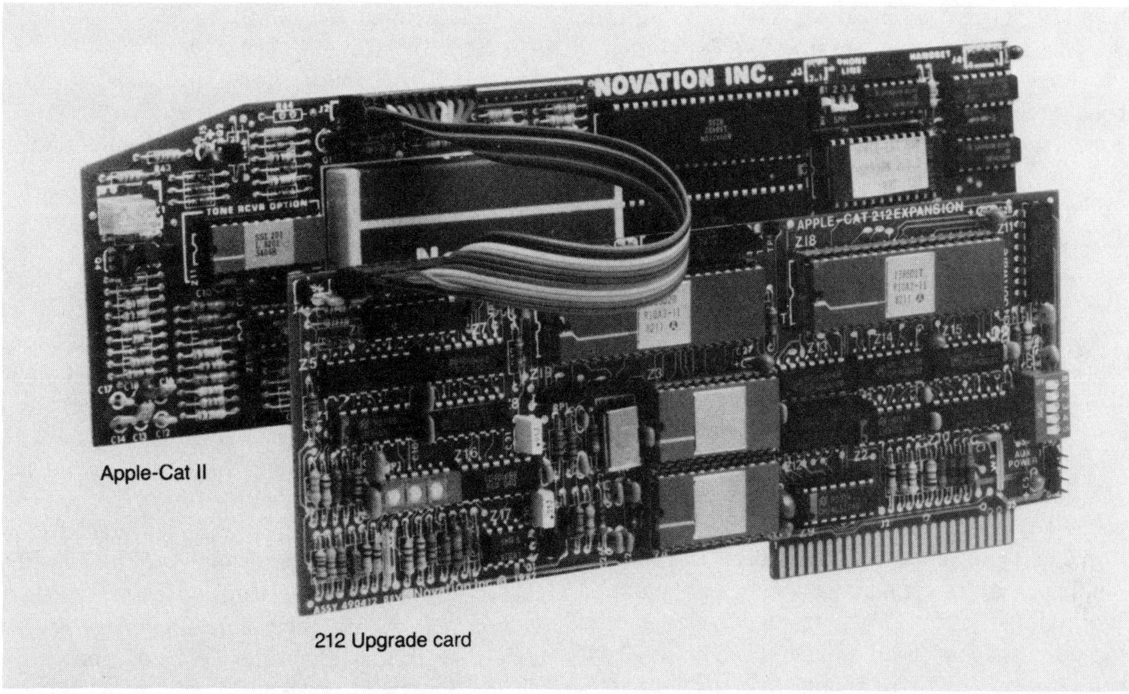

Fig. 8-28. The Apple Cat II. Courtesy Novation Inc.

- A synchronous switch for communicating with 212 systems in synchronous mode.
- A remote digital loop in the receiver allows another modem to execute a digital loopback test.
- A send space disconnect transmits 4 seconds of spacing to a remote computer, which causes some remote modems to go on hook.
- A receive space disconnect will disconnect a data call after receiving approximately 1.6 seconds of continuous spacing.
- Loss of carrier disconnect terminates a data call when the received carrier is lost for 350 milliseconds.
- An external clock switch allows an external clock to determine the timing signal of the modem.
- A slave clock provides internal timing for the modem.
- Two switches determine the word length of transmitted characters as 8, 9, 10, or 11 bits.
- 1200 baud sets the modem for high speed operation.

In addition to these 10 switches, 7 front panel switches are used for routine functions. They include: an originate switch to connect the modem after manually dialing; an answer switch, to automatically answer an incoming call; four switches for various self-tests; and a high-speed switch for 1200-baud operation.

Apple-Cat II

BAUD: 1200/300
COMPUTER: Apple II, Apple II+
TYPE: Bell 103/202, with upgrade card Bell 212
SOFTWARE: Included (Com Ware II)
AUTO ANSWER: Yes
AUTO DIAL: Yes (pulse- or tone-dial)
PHONE: RJ11 (modular) connection

Features. The Apple Cat is an internal modem that inserts directly into available slots within the Apple II or Apple II+ (Fig. 8-28). It requires 48K of memory, at least one disk drive, and a 3.2, 3.2.1, or 3.3 Apple disk operating system.

The modem operates in a Bell 103 mode at baud rates of 0 to 300 and in a Bell 202 mode at 1200 baud. An upgrade card that connects directly to the modem permits Bell 212A compatibility. Additional options enable the Apple-Cat II to interface with BSR modules to control appliances by remote control, and an interface to control a cassette recorder. With special programming, the recorder can be used to record incoming calls on an unattended system. Additional features include auto-dial and auto-answer.

Discussion. The software included with Apple-Cat II displays all of the functions in a selectable list. Various features are selected by pressing a single key referenced on the list.

The upper portion of the screen displays the status of the system during all modes of operation both on and off line. Text or binary files can be uploaded or downloaded through a 27K buffer. Text can be sent directly to a serial or parallel printer. Phone numbers can be dialed through the keyboard or from a list of 26 numbers created and saved to diskette.

PENRIL DATACOMM

207 Perry Parkway
Gaithersburg, MD 20877
(301) 921-8600

The Data Communications Division of Penril warrants that equipment manufactured and sold will be free from defects in materials and workmanship for a period of 1 year. Defective equipment can be returned to the technical-support department for service or replacement.

Models reviewed and suggested retail prices:

8216 (Auto Data 300/1200)	$525.00
8216 with auto-dial	$650.00

Auto Data 300/1200

BAUD: 0-300,1200
COMPUTER: Any RS-232C-compatible system
TYPE: Bell 103,113,212A
SOFTWARE: Not included
AUTO ANSWER: Yes
AUTO DIAL: No
PHONE: RJ11C (modular) connection or others by request

Features. The Auto Data 300/1200 provides full-duplex operation on two-wire switched networks (Fig. 8-29). It operates at 0 to 300 baud, asynchronously, using a Bell 103 or 113 format, or at 1200 baud, synchronously or asynchronously, in a Bell 212 format. It comes in a stand-alone case or as a single board for rack mounting, using an optional adapter.

The front-panel display includes eight LEDs which indicate the operating status of the modem. These are: make busy, terminal ready, modem ready, transmit data, receive data, high/low speed, modem check, and test mode. Test mode includes a self-test, analog loopback test, remote digital loopback test, and local digital loopback test. Buttons on the front panel implement the various tests and select between data and talk modes.

The rear panel includes connections for power, an RS-232C connection, an RJ11C telco connection to an appropriate wall jack, and RJ11C tel-set connection to a compatible phone, and a button to select between high-speed and low-speed operations.

The modem can be configured for various modes of operation through strap settings located inside it. The documentation includes instructions on how to access these settings and make the appropriate changes.

Discussion. The Universal Data System 212A uses frequency-shift keying at 0 to 300 baud or differential phase-shift keying at 1200 baud. The various parameters selected through internal switch settings include a speed control, which enables the modem to automatically adjust its speed or allow manual control; timing control, which allows timing to be controlled internally or externally; 1200 bps operation, which establishes synchronous or asynchronous operation; character length of 9 or 10

Fig. 8-29. The Auto Data 300/1200 modem. Courtesy Penril DataComm.

bits, including stop bits; loss of carrier disconnect for automatic disconnect if a loss-of-carrier signal occurs; disconnect on received space, for disconnects if long steady space is received; automatic answer; receiver response to digital loop, which allows your modem to automatically set itself for a digital loop-back test from a remote modem; and a signal ground to frame, which ties the signal ground to the frame ground.

The buttons on the front panel allow the user to select between data mode and talk mode. In the talk mode, the phone essentially separates from the modem circuit for normal voice communication. In the data mode, the phone is used in conjuction with the modem for data transmission.

The receive-data LED indicates that data has been received by the modem and passed to the terminal equipment. The transmit LED indicates that data has been received from the terminal. The make-busy LED indicates the modem is operating in the analog-loopback mode. The terminal-ready LED indicates that the data-terminal-ready line from the terminal is active. The modem-ready LED indicates the modem is ready to transmit or receive data. The high/low LED indicates the modem is operating at 1200 baud. The modem-check LED indicates errors during a self-test. The test-mode LED is on during any modem test.

The analog loopback test isolates the modem from the telephone line and loops the transmit signal back through the receive circuitry for display on your video display unit. The self-test can be used with the digital tests to transmit a test pattern and to check received data for errors.

The remote digital loopback signals the modem at the other end of the line to enter the digital-loopback mode. Characters transmitted from your terminal will be looped back from the distant modem

Fig. 8-30. The 8216 with Auto Dial. Courtesy Penril Data Comm.

for verification. The digital loopback takes data received over the phone lines and loops it back over the lines to the originating system.

8216 with Auto Dial

BAUD: 0-300,1200
COMPUTER: Any RS-232C-compatible system
TYPE: Bell 103,113,212A
SOFTWARE: Not included
AUTO ANSWER: Yes
AUTO DIAL: Yes (pulse- or tone-dial)
PHONE: RJ11C (modular) connection or others by request

Features. The 8216 with Auto Dial provides full-duplex operation on two-wire switched networks (Fig. 8-30). It operates asynchronously at 0-300 baud, using a Bell 103 or 113 format, or at 1200 baud, synchronously or asynchronously, using a Bell 212 format. It can store up to 10 numbers, which can be automatically dialed from single keyboard entries. Each number can include commands for various dialing activities, such as redialing.

The display on the front panel includes eight LEDs which indicate the operating status of the modem. They are: make busy, terminal ready, modem ready, transmit data, receive data, high/low speed, modem check, and test mode. The test mode includes a self-test, analog loopback test, remote digital loopback test, and local digital loopback test. Buttons on the front panel implement the various tests and select between data and talk modes.

The rear panel includes connections for power, an RS-232C connection, an RJ11C telco connection to an appropriate wall jack, an RJ11C tel-set connection to a compatible phone, and a button to select between high- or low-speed operation.

The modem can be configured for various modes of operation through strap settings located inside the modem. The documentation includes instructions on how to access these settings and make the appropriate changes.

Discussion. The modem uses frequency-shift keying at 0 to 300 baud, or differential phase-shift keying at 1200 baud.

The various parameters selected through internal switch settings include a speed control, timing control, 1200 bps operation, character length, loss of carrier-disconnect, disconnect on received space, automatic answer, receiver response to digital loop, and a signal ground to frame.

The 8216 with auto dial recognizes a series of one- or two-character commands from the keyboard which allow you to dial numbers from the keyboard or to store up to 10 numbers in battery-protected memory inside the modem. Each number can consist of up to 20 characters, including some command characters. The commands include "R" (redial), "U" (dial until answered or 15 busy signals), "C" (clear a number from memory), "D" (display the contents of the modems memory), and "H" (to view a list of commands). Additional commands select between pulse- or touch-tone dialing, insert descriptive text in memory, and perform various tests. A special link command will automatically dial another number if the first is busy or does not answer.

The buttons on the front panel allow the user to select between data mode and talk mode. In the talk mode, the phone is essentially removed from the modem circuit and is used normally for voice communication. In the data mode, the phone is used in conjunction with the modem for data transmission.

The receive-data LED indicates that data has been received by the modem and passed to the terminal equipment. The transmit LED indicates that data has been received from the terminal. The make-busy LED indicates the modem is operating in the analog-loopback mode. The terminal-ready LED indicates that the data-terminal-ready line from the terminal is active. The modem-ready LED indicates the modem is ready to transmit or receive data. The high/low LED indicates the modem is operating at 1200 baud. The modem-check LED indicates errors during a self-test. The test-mode LED is on during any modem test.

The analog loopback test isolates the modem from the telephone line and loops the transmit signal back through the receive circuitry for display on your video display unit. The self-test can be used with the digital tests to transmit a test pattern and to check received data for errors.

The remote digital loopback signals the modem at the other end of the line to enter the digital loopback mode. Characters transmitted from your terminal will be looped back from the distant modem for verification. The digital loopback takes data received over the phone lines and loops it back over the lines to the originating system.

PROMETHEUS PRODUCTS, INC.

45277 Fremont Blvd.
Fremont, CA 94538
(415) 490-2370

Prometheus warrants its products to be free from defects due to faulty manufacture of the product or its components for a period of 1 year from the date of sale.

Models reviewed and suggested retail prices:

ProModem 1200 $495.00

ProModem 1200

BAUD: 1200/300
COMPUTER: Any RS-232C-compatible system
TYPE: Bell 212A/103/113
SOFTWARE: Not included
AUTO ANSWER: Yes
AUTO DIAL: Yes (pulse- or tone-dial)
PHONE: RJ11 (modular) connection

Features. The ProModem is Bell 212A- and Bell 103-compatible with operating speeds of 1200 baud and 0 through 300 baud (Fig. 8-31). It incorporates a microprocessor and associated firmware to provide auto-dial and auto-answer capabilities. The modem may be used in conjunction with most communications programs using Hayes modem commands.

The ProModem includes a built-in real-time clock/calendar. The clock may be accessed under program control for use in applications programs, or in conjunction with modem commands to execute commands at programmed times.

The ProModem has two modular phone (RJ11C) sockets, one for connection to an appropriate phone outlet and one for connection to a modular phone. A female DB25 connector in the modem permits connection to any standard RS-232C interface with a like connector. It is presumed that a connected computer will have appropriate software to implement the RS-232C standard.

Dialing commands enable either pulse-or-tone dialing, automatic redialing, and delays between numbers (when a second dial tone exists). Auto-answer commands can set the modem to answer the phone after a predetermined number of rings.

A series of 10 switches facilitate configuration of the modem for specific applications. Eight LEDs monitor the status of the modem.

Fig. 8-31. The ProModem 1200. Courtesy Prometheus Products, Inc.

Result codes which, monitor the progress of calls, such as CONNECT, BUSY, DIALING, etc., can be displayed on your terminal's video display unit. An optional processor, memory expansion, and digit alphanumeric display are available to enhance capabilities.

Discussion. The modem assumes one of two states: the command mode or an on-line mode. The modem assumes the command state when first powered up. One- or two-character commands are entered from the keyboard. The commands are used to set RS-232C parameters, such as baud rate and duplex, and also to implement the various dialing instructions. The on-line mode is used to transfer data to or from the modem.

The modem will automatically disconnect if it is unable to obtain a carrier frequency from a remote modem after a programmed amount of time. Once obtained, if the carrier frequency is lost for more than a programmed amount of time, the modem will automatically disconnect.

LEDs on the front panel indicate various conditions:

- The modem is ready to operate.
- A valid carrier is detected from a distant modem.
- Data is being transmitted.
- Data is being received.
- The modem is operating at 1200 baud.
- The modem is in the auto-answer mode.
- The phone is off the hook.
- There is an error in diagnostics.

A panel switch activates command recognition when you are planning to transmit commands to the modem. It disables command recognition when you are using the modem only for answering calls. Another switch activates a single-line connection, when using RJ11C connections, or a multi-line connection, when using other types of lines. Switch settings select between Promodem result codes and Hayes result codes, in either numeric or English format. A switch setting configures the modem for auto-answer. Two switch settings allow you to disable the data-terminal-ready line or carrier-detect line, to interact with certain computers. A switch lets the ProModem redial a number automatically every 30 seconds if the number is busy.

An automatic self-test verifies the working condition of RAM and ROM memories, as well as other critical components. The documentation includes notes to enable you to configure the modem to work with most communication programs which use the Hayes command set, including Crosstalk, Perfect Link, Smartcom, and Zip-Comm.

RACAL-VADIC

222 Caspian Dr.
Sunnyvale, CA 94089-9874
(408) 744-0810

Racal-Vadic is a pioneer in communications technology, having developed the first full-duplex, 1200-baud modem several years ago. It was one of the first to go up against AT&T's virtual monopoly on communications hardware.

Specializing in communications hardware, Racal-Vadic offers standard modems, as well as custom modems for special needs. Racal-Vadic warrants its products for 1 year from the date of purchase. Its regional service centers answer questions and provide diagnostic tests via the phone.

Models reviewed and suggested retail prices:

Maxwell 300 PC	$325.00
Maxwell 1200 PC	$595.00
Maxwell 300V	$350.00
Maxwell 1200V	$595.00
VA1222	$975.00
Auto Dial VA212	$695.00
VA212 LC	$495.00
Auto Dial VA3451	$875.00

Maxwell 300 PC

BAUD: 300
COMPUTER: IBM PC, IBM XT, IBM portable, IBM-compatible
TYPE: Bell 103/113
SOFTWARE: Included (George)
AUTO ANSWER: Yes
AUTO DIAL: Yes (pulse- or tone-dial)
PHONE: RJ11 (modular) connection

Features. The Maxwell 300 PC is an internal modem that inserts into the half-sized slot in the IBM PC or IBM XT, leaving the full-size slots available for expansion (Fig. 8-32). Commands provide for touch-tone dialing, pulse-dialing, and repeat dialing instructions. The modem also includes an auto-answer feature, allowing callers to establish contact

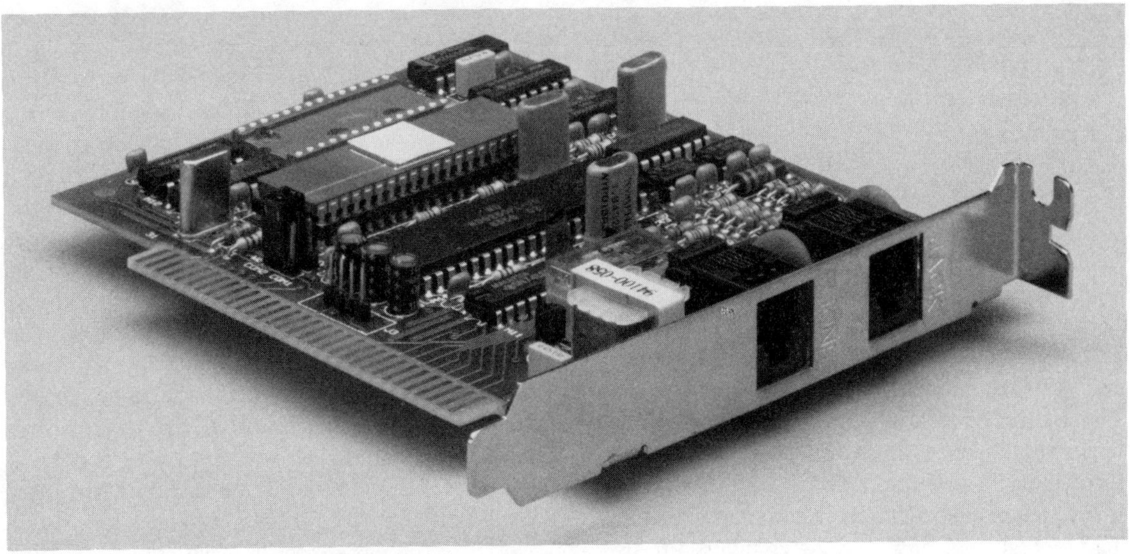

Fig. 8-32. The Maxwell Modem 300 PC. Courtesy Racal-Vadic.

and leave messages on an unattended system.

The modem can be configured to recognize Racal-Vadic dialing commands, used by the software *George*, or the Hayes dialing commands, used by a number of other programs.

The modem includes circuitry for both analog and digital loopback tests. Racal-Vadic provides a remote diagnostic service which can be used to troubleshoot problems with a Racal-Vadic modem.

Discussion. Many of the functions of the modem actually come from the software included with the modem. A directory of 24 entries—each with its own dialing operators, sign-on sequences, and format instructions—can be created and used to connect with remote services automatically. Text and binary files can be downloaded or uploaded. XMODEM protocol can be used for complete error detection. Special keys can be programmed to provide macro instructions, allowing connection to another system with a single keystroke. A message editor facilitates the creating and editing of short messages. A status line on the screen provides current feedback on the status of the program.

A special dialing feature allows tandem dialing, where two or more sequences must be dialed in succession. For use with MCI, Sprint, and PBX switchboards, the modem will dial a series of numbers, "listen" for a new dial tone, and proceed to dial the next series of numbers. For systems with unusual dial tones, the modem can be configured to ignore dial tones and dial after a programmed delay.

A series of programmable options can be altered through keyboard commands. The options include: carrier control, DSR control, auto-answer, dial mode, blind dial, response mode, local copy, character length, disconnect control, and call progress. These options assist in configuring the modem for various communications programs and applications.

Maxwell modem is designed to recognize one- and two-character commands from the keyboard. These commands configure the modem and execute dialing and answering features. The characters used for commands are determined by Racal-Vadic, and many are automatically used in the program George. The modem can, however, be configured to use a subset of Hayes commands, used by some other communications software, making the Maxwell modems compatible with most programs using Hayes commands.

Maxwell 1200 PC

BAUD: 1200/300
COMPUTER: IBM PC, IBM XT, IBM portable, IBM-compatible
TYPE: Bell 212A/103/113
SOFTWARE: Included (George)
AUTO ANSWER: Yes
AUTO DIAL: yes (pulse- or tone-dial)
PHONE: RJ11 (modular) connection

Features. The Maxwell 1200 PC is an internal modem that inserts into a full-size slot of the IBM PC or IBM XT. The modem operates at 1200 baud in a Bell 212A format, or at 0 to 300 baud in a Bell 103 format.

Commands provide for touch-tone dialing, pulse-dialing, and repeat dialing instructions. The modem also includes an auto-answer feature, allowing callers to establish contact and leave messages on an unattended system.

The modem can be configured to recognize Racal-Vadic dialing commands, used by the software George, or the Hayes dialing commands, used by a number of other programs.

The modem includes circuitry for both analog and digital loopback tests. Racal-Vadic provides a remote diagnostic service which can be used to troubleshoot problems with a Racal-Vadic modem.

Discussion. Many of the functions of the modem actually comes from the software George, included with the modem. A directory of 24 entries—each with its own dialing operators, sign-on sequences, and format instructions—can be created and used to automatically connect with remote services. Text and binary files can be downloaded or uploaded. XMODEM protocol can be used for complete error detection. Special keys can be programmed to provide macro instructions, allowing connection to another system with a single keystroke. A message editor facilitates the creating and editing of short messages. A status line on the screen provides current feedback on the status of the program.

A special dialing feature allows tandem dialing, where two or more sequences must be dialed in succession. For use with MCI, Sprint, and PBX switchboards, the modem will dial a series of numbers, "listen" for a new dial tone, and proceed to dial the next series of numbers. For systems with unusual dial tones, the modem can be configured to ignore dial tones and dial after a programmed delay.

A series of programmable options can be altered through keyboard commands. The options include: carrier control, DSR control, auto-answer, dial mode, blind dial, response mode, local copy, character length, disconnect control, and call progress. These options assist in configuring the modem for various communications programs and applications.

The Maxwell modem is designed to recognize one- and two-character commands from the keyboard. These commands configure the modem and execute dialing and answering features. The characters used for commands are determined by Racal-Vadic, and many are automatically used in the program George. The modem can, however, be configured to use a subset of Hayes commands, used by some other communications software, making the Maxwell modems compatible with most programs using Hayes commands.

Maxwell 300V

BAUD: 300
COMPUTER: Any RS-232C-compatible systems
TYPE: Bell 103/113
SOFTWARE: Not included
AUTO ANSWER: Yes
AUTO DIAL: Yes (pulse- or tone-dial)
PHONE: RJ11 (modular) connection

Features. The Maxwell 300V is an external 300-baud modem that connects through a standard DB25 connector to RS-232C-compatible systems (Fig. 8-33). Commands provide for touch-tone dialing, pulse-dialing, and repeat dialing instructions. The modem also includes an auto-answer feature, allowing callers to establish contact and leave messages on an unattended system.

The modem can be configured to recognize Racal-Vadic dialing commands, used by the software George, or the Hayes dialing commands, used by a

Fig. 8-33. The Maxwell 300V. Courtesy Racal-Vadic.

number of other programs.

The modem includes circuitry for both analog and digital loopback tests. Racal-Vadic provides a remote diagnostic service which can be used to troubleshoot problems with any of its modems.

Discussion. A special dialing feature allows tandem dialing where two or more sequences must be dialed in succession. For use with MCI, Sprint, and PBX switchboards, the modem will dial a series of numbers, "listen" for a new dial tone, and proceed to dial the next series of numbers. For systems with unusual dial tones, the modem can be configured to ignore dial tones and dial after a programmed delay.

A series of programmable options can be altered through keyboard commands. The options include: carrier control, DSR control, auto answer, dial mode, blind dial, response mode, local copy, character length, disconnect control, and call progress. These options assist in configuring the modem for various communications programs and applications.

The modem is designed to recognize one- and two-character commands from the keyboard. These commands configure the modem and execute dialing and answering features. The characters used for commands are determined by Racal-Vadic, and many are automatically used in the program George, available from Racal-Vadic. The modem can also be configured to use a subset of Hayes commands, used by some other vendors of communications software.

Maxwell 1200V

BAUD: 1200/300
COMPUTER: Any RS-232C-compatible system
TYPE: Bell 212A/103/113
SOFTWARE: Not incuded
AUTO ANSWER: Yes
AUTO DIAL: Yes (pulse- or tone-dial)
PHONE: RJ11 (modular) connection

Features. The Maxwell 1200V is an external modem that connects through a standard DB25 connector to RS-232C-compatible systems (Fig. 8-34). The modem operates at 1200 baud in a Bell 212A format, or 0-300 baud in a Bell 103 format.

Commands provide for touch-tone dialing, pulse-dialing, and repeat dialing instructions. The modem also includes an auto-answer feature, allowing callers to establish contact and leave messages on an unattended system.

The modem can be configured to recognize Racal-Vadic dialing commands used by the software George, or the Hayes dialing commands, used by a number of other programs.

The modem includes circuitry for both analog and digital loopback tests. Racal-Vadic provides a remote diagnostic service to troubleshoot problems.

Discussion. A special dialing feature allows tandem dialing where two or more sequences must be dialed in succession. For use with MCI, Sprint, and PBX switchboards, the modem will dial a series of numbers, "listen" for a new dial tone, and proceed to dial the next series of numbers. For systems with unusual dial tones, the modem can be configured to ignore dial tones and dial after a programmed delay.

A series of programmable options can be altered through keyboard commands. The options include: carrier control, DSR control, auto answer, dial mode, blind dial, response mode, local copy, character length, disconnect control, and call progress. These options assist in configuring the modem for various communications programs and applications.

The Maxwell modem is designed to recognize one- and two-character commands from the keyboard. These commands configure the modem and execute dialing and answering features. The characters used for commands are determined by Racal-Vadic, and many are automatically used in the program George. The modem can also be configured to use a subset of Hayes commands, used by some other communications software.

VA1222

BAUD: 1200/600/0-300
COMPUTER: RS-232C and V.24/V.28-compatible systems
TYPE: CCITT V.22, V.24, V.25, V.28, V.54
SOFTWARE: Not included

Fig. 8-34. The Maxwell 1200V, shown with the IBM PC. Courtesy Racal-Vadic.

AUTO ANSWER: Yes
AUTO DIAL: No
PHONE: RJ11 (modular) connection

Features. The VA1222 is a full-duplex 1200-baud modem designed to operate over public or private two-wire lines. The modem is compatible with the international standards just listed and is intended for accessing information for foreign host systems. The modem comes as a stand-alone model or with a rack mount capable of accommodating seven modem cards.

The modem includes an auto-answer feature for answering calls on an unattended system. An automatic-calling unit can be purchased separately.

The modem includes circuitry for both analog and digital loopback tests. Racal-Vadic provides a remote diagnostic service which can be used to troubleshoot problems.

Discussion. Seven front-panel switches control various features of the modem. One switch configures the modem for use on public or private line systems. A second switch implements the auto-answer mode. A third switch selects the data-transmission frequency, to avoid conflicts in full-duplex exchanges. Four more switches execute the various analog and digital loopback tests.

Eight LEDs provide information on the status of the modem:

- Data is being transmitted.
- Data is being received.
- The connected terminal or computer is ready.
- The modem itself is ready.

- The modem is detecting a carrier signal from the other modem.
- Proper voltage has been applied.
- There are errors in testing.
- Testing is occurring.

Various modes involving selected baud rates are implemented through a series of jumpers located inside the modem.

Auto Dial VA212

BAUD: 1200/300
COMPUTER: Any RS-232C-compatible system
TYPE: Bell 103/212
SOFTWARE: Not included
AUTO ANSWER: Yes
AUTO DIAL: Yes (pulse- or tone-dial)
PHONE: RJ11 (modular) connection

Features. The Auto Dial VA212 is a Bell 103- and Bell 212A-compatible modem (Fig. 8-35). Some of its features include automatic dialing, automatic selecting of tone- or pulse-dialing, automatic answering, and redialing of the last number for a specified number of times.

Features and options can be executed through software control by entering special characters through the keyboard of the computer or directly from the front panel of the modem. Current modes of operation are displayed on both the computer's video display and the liquid-crystal display (LCD) of the modem.

The VA212 has extensive user diagnostics, and local as well as remote self-tests. A unique password of up to six alphanumeric characters can be programmed into the modem's memory. Access to any of the stored numbers requires entry of the programmed password. Two RJ11C jacks provide connections to a telephone wall jack and telephone.

Discussion. The Auto Dial VA212 modem responds to single-character commands that are entered through the keyboard of the computer. For example, "S' stores a phone number; "M" displays previously stored numbers; "C" changes previous numbers, and "L" allows you to link any two numbers in memory to automatically dial the second number if the first number is busy.

Prior to dialing, the modem recognizes and indicates a dial tone. It recognizes when the phone at

Fig. 8-35. The Auto Dial VA212 modem. Courtesy Racal-Vadic.

Fig. 8-36. The VA212 LC modem. Courtesy Racal-Vadic.

the other end is ringing and displays "ringing" on the screen. If busy, the modem will display "No Dial Tone." If the call gets a noncomputer response, the screen displays "Voice."

The VA212 features 26 user-programmable options to accommodate special applications. These options include the settings for word length, baud rate, originate or answer mode, and dialing mode. Certain lines used in RS-232C communications can be forced on or off, such as those for data set ready, carrier detect, and data terminal ready. All settings can be stored in memory for later use.

Diagnostics include a local analog test that will identify problems between your computer and the modem, and a remote digital loopback test that will locate problems between your computer and the modem at the other end of transmission.

VA212 LC

BAUD: 1200/300
COMPUTER: Any RS-232C-compatible system
TYPE: Bell 103/212
SOFTWARE: Not included
AUTO ANSWER: Yes
AUTO DIAL: No
PHONE: RJ11 (modular) connection

Features. The VA212 LC is a dual-speed, Bell 103- and Bell 212A-compatible modem (Fig. 8-36). Its features include a manual-dial mode, and an automatic-answer mode.

The VA212 LC has an automatic self-test that activates when the modem is not in use, and a remote test that allows a regional service center to run diagnostics on the modem over the phone. This modem provides two RJ11C jacks, permitting use of the phone for either data or normal voice communications.

Discussion. The VA212 LC has two modes of operation: manual orginate, for placing calls to another modem; and automatic answer, for receiving calls from another modem. In the manual originate mode, you dial numbers from the phone manually. On establishing contact, you press a data switch on the front panel to allow data transmission. At the end of transmission, you press a VOICE switch on the front panel to disconnect the modem. If the remote terminal fails to respond with a valid carrier signal, the modem automatically disconnects within 20 seconds. If the incoming transmission interrupts, the modem disconnects within 1 second.

Automatic-answer preparation comes from leaving the modem in the data mode. LEDs on the front panel give some indication of the progress of an incoming call. The modem will disconnect if the incoming call does not provide a carrier signal from a remote modem within 20 seconds. It will disconnect within 1 second if transmission is interrupted from the remote modem. It can be manually disconnected by pressing the VOICE mode switch on the front panel.

The front panel has four buttons: VOICE, to use the phone for normal conversation; DATA, to use the phone for data transmission; HIGH SPEED, for 1200-baud operation; and LOW SPEED, for 300-baud operation.

Five LEDs on the front panel give some indication of the status of the modem:

- Data is being transmitted.
- Data is being received.
- The modem is receiving a valid signal from a remote modem.
- The modem is in data mode.
- It is operating at 1200 baud.

A self-test runs automatically when the modem is not in use. It indicates an error by illuminating the carrier-detect LED. A regional service center can run diagnostics on the VA212 LC through the connected phone.

Auto Dial VA3451

BAUD: 1200/300
COMPUTER: Any RS-232C-compatible system
TYPE: Bell 103/212, Racal Vadic VA 3400
SOFTWARE: Not included
AUTO ANSWER: Yes
AUTO DIAL: Yes
PHONE: RJ11 (modular) connection

Features. The VA3451 can emulate Bell 103-, Bell 212A-, or VA3400-type modems (Fig. 8-37). It features automatic dialing and answering. The modem can automatically dial telephone numbers, either directly from the keyboard, or from up to 9 entries stored in memory. If the dialed number is busy, the modem will automatically redial it a specified number of times. The modem automatically adjusts to the type of calling modem.

Auto-dial and auto-answer features are programmed from the keyboard of the computer with single-character commands. The screen displays progress messages. Additional options are implemented through switches and solder straps inside the modem.

The VA3451 has extensive diagnostics, including local and remote, analog and digital loopback tests. A single RJ11C outlet is connected to a standard wall outlet. A special T-adaptor is included, allowing a simultaneous connection of a telephone for normal use when not transmitting data.

Discussion. The VA3451 modem recognizes single-character commands that can be sent from the keyboard of the computer: S, to store a phone number; M, to display previously stored numbers; R, to redial a previous number; and M, to display a number in memory.

The VA3451 features numerous user-programmable options to accommodate special operating requirements of certain installations. These options are set with DIP switches and straps inside the modem. They include the settings for

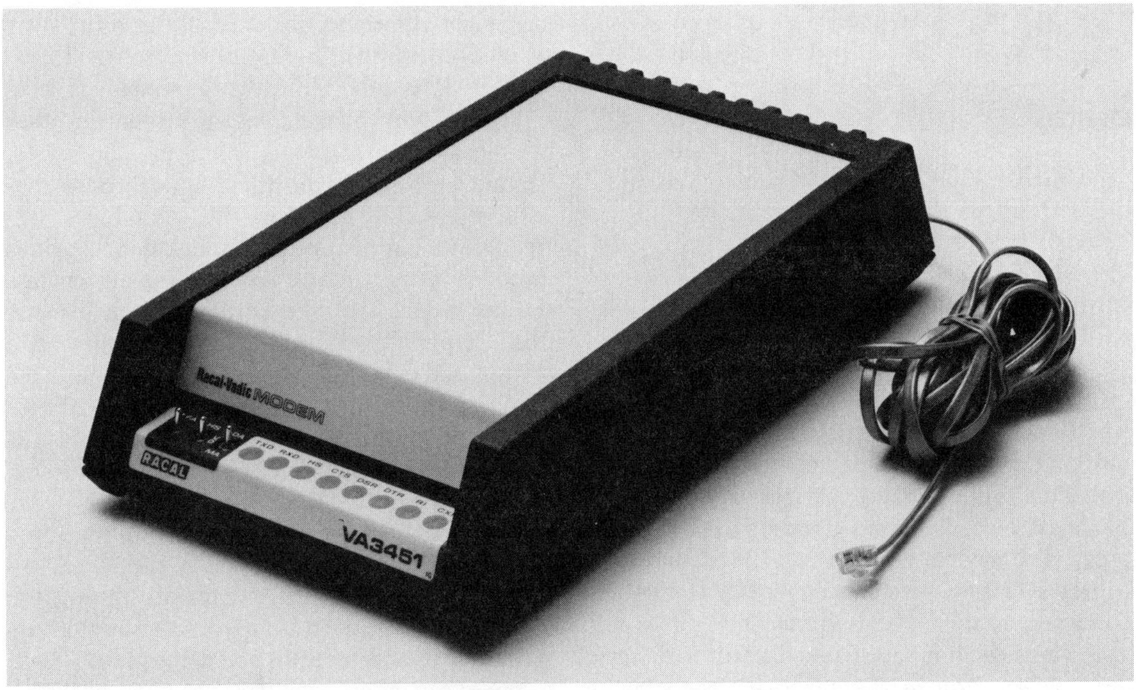

Fig. 8-37. The VA3451 modem. Courtesy Racal-Vadic.

word length, baud rate, originate or answer mode, and dialing mode. Several RS-232C lines can be forced on or off, such as those for data set ready, carrier detect, and data terminal ready. The modem can be set to disconnect after losing a carrier signal or after receiving a series of space characters from the remote modem. Timing controls can be triggered internally or externally. The transmit levels can be programmed with special resistors.

Eight LEDs indicate the current operating status of the modem:

- Transmitting data.
- Receiving data.
- Operating at 1200 baud.
- Modem ready to transmit data.
- Terminal ready to transmit.
- Detecting a ring voltage.
- Receiving a valid carrier signal from a remote modem.
- Modem connected.

Front-panel switches select between a manual-answer mode, an automatic-answer mode, and data mode. Additional switches select between 1200- or 300-baud operation.

The modem has a number of diagnostic routines for locating problems with communication. A local analog test will identify problems between your computer and the modem. A remote digital loopback test will locate problems between your computer and the modem at the other end of transmission.

RIXON, INC.

2120 Industrial Parkway
Silver Spring, MD 20904
(301) 622-2121

Rixon warrants to the original buyer for 2 years from the date of purchase that, under normal use and except for batteries, its products will be free of defects in materials and workmanship. Products can be shipped to Rixon, Inc., for repair or service.

Models reviewed and suggested retail prices:

PC212A	$539.00
R212A	$899.00

PC212A

BAUD: 1200/300 Asynchronous
COMPUTER: IBM PC
TYPE: Bell 212A/103/113
SOFTWARE: Included (PC COM I)
AUTO ANSWER: Yes
AUTO DIAL: Yes (pulse- or tone-dial)
PHONE: RJ11C (modular) connection

Features. The PC212A modem is an internal modem that inserts into any of the full-size expansion slots (Fig. 8-38). It features an integral automatic dialer and two asynchronous serial interface ports. The modem includes two RJ11C jacks that allow you to leave a telephone connected for normal use when the computer is not on line.

The modem includes auto-dial features, using either pulse- or tone-dial systems. It can store up to 10 phone numbers of 60 characters each in battery-protected memory. Only a single keystroke is necessary to dial or display a number. Numbers can be dialed from the keyboard or from PC COM I software. If the first number is busy, a linking feature can automatically dial a second number.

The PC212A supports auto-answer features, implemented through software, and monitors the progress of calls with messages, such as WAITING, RINGING, NO ANSWER, etc. This modem recognizes the Hayes command set and works with other communications software, including Crosstalk. The modem can institute an analog loopback test, to verify its working condition.

Discussion. The documentation provides thorough instructions for installing the modem into the computer. The modem features an external RS-232C port for additional serial devices, such as a printer, and two RJ11C ports for connection to the phone system. Internal switch settings configure the device as Com Port 1 or Com Port 2. Many of the functions of the modem are actually the result of the software PC COM I. (See Chapter 9.)

The modem can be operated to recognize either the Rixon command set or the Hayes abbreviated Smartmodem commands. Using the Rixon commands, up to 10 numbers can be stored in the modem's battery-protected memory. These numbers can include commands to redial a number or to dial a different number if the first number is busy. The modem can be programmed to wait for secondary dial tones, for dialing through PBX systems. Calls can also be made directly from the keyboard, or from a directory of numbers, using PC COM I.

Commands for auto-dial provide for touch-tone

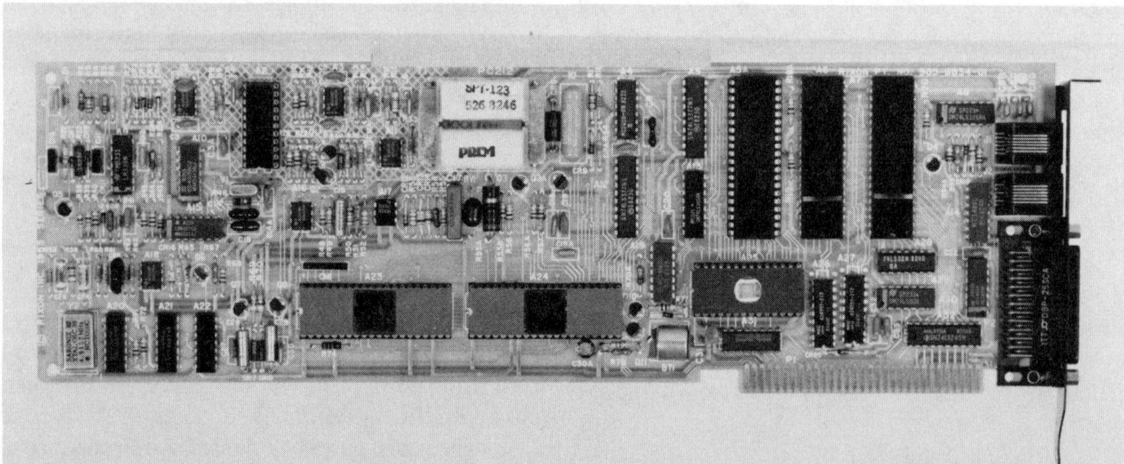

Fig. 8-38. The PC212A modem. Courtesy Rixon, Inc.

Fig. 8-39. The R212A. Courtesy Rixon, Inc.

dialing, pulse-dialing, pausing between numbers, repeating dialing instructions, dialing secondary numbers, displaying stored numbers, erasing stored numbers, and adding text for descriptions of numbers. Normally, a number is not dialed until the modem senses a dial tone. The PC212A has the ability to *blind dial*, or dial without the dial tone, for phones with nonstandard tones.

A programmed disconnect allows the manual disconnection of the modem from the line, essentially allowing you to hang up the phone in an orderly fashion. The modem has auto-answer facilities that are essentially implemented through the software.

An analog loopback test is executed through a single keyboard command. Following the command, any characters entered from the keyboard are transmitted to the modem and returned back for display. If the displayed characters are incorrect, a problem exists with the modem, computer, or connection.

The Hayes command set can be directed either from the keyboard or from program control. Hayes commands cannot, however, be used from the modem's memory.

R212A

BAUD: 1200/300 asynchronous
COMPUTER: Any RS-232C-compatible systems
TYPE: Bell 212A / 103 / 113
SOFTWARE: Not included
AUTO ANSWER: Yes
AUTO DIAL: Yes (pulse- or tone-dial)
PHONE: RJ11C (modular) connection

Features. The R212A modem is an external modem that connects to most computers through a standard RS-232C type interface (Fig. 8-39). The modem includes two RJ11C jacks that allow you to leave a telephone connected for normal use when the computer is not on line. The panel has no switches; all activities are controlled through keyboard commands.

The modem includes auto-dial features, using either pulse-dial or tone-dial systems. It can store up to 10 phone numbers of up to 60 characters each in battery-protected memory. Only a single keystroke is necessary to dial or display a number. If the first number is busy, a linking feature can automatically dial a second number.

The R212A supports auto-answer features, implemented through software, and monitors the progress of calls with messages such as WAITING, RINGING, NO ANSWER, etc. This modem recognizes the Hayes command set and works with other communications software, including PC COM I and Crosstalk. It can institute an analog loopback test, to verify its working condition.

Discussion. The documentation provides

thorough instructions for connecting the modem to the computer. It provides an external RS-232C port for connecting a standard DB25 cable, and two RJ11C ports for connecting to the phone system.

The modem can be operated to recognize either the Rixon command set or the Hayes abbreviated Smartmodem commands. Using the Rixon commands, up to 10 numbers can be stored in the modem's battery-protected memory, including commands to redial a number or to dial a different number if the first number is busy. Delays can be inserted to wait for secondary dial tones, for dialing through PBX systems. Calls can also be made directly, from the keyboard, or through software options, provided by various communications programs, such as PC COM I.

Commands for auto-dial provide for touch-tone dialing, pulse-dialing, pausing between numbers, repeating dialing instructions, dialing secondary numbers, displaying stored numbers, erasing stored numbers, and adding text for descriptions of numbers. Normally, a number is not dialed until the modem senses a dial tone. The R212A has the ability to blind dial, so it can be used on phones with nonstandard tones.

A programmed disconnect allows the manual disconnection of the modem from the line, essentially allowing you to hang up the phone in an orderly fashion. The modem has auto-answer facilities that are essentially implemented through the software.

An analog loopback test is executed through a single keyboard command. Following the command, any characters entered from the keyboard are transmitted to the modem and returned back for display. If the displayed characters are incorrect, a problem exists with the modem, computer, or connection.

The Hayes command set can be directed from the keyboard or from program control. Hayes commands cannot, however, be implemented from the modem's memory.

TEK-COM CORPORATION

1887-C 103 O'Toole Ave.
San Jose, CA 95131
(408) 263-7400

Tek-Com Corporation is a wholly owned subsidiary of Prentice Corporation. Tek-Com markets a number of acoustic modems, Bell 103-compatible modems, Bell 202S-compatible modems, Bell 212A-compatible modems, limited-distance, high-speed modems (0-9600 baud), and accessory cables. Products are warranted to be free from defects in parts and workmanship for 1 year.

Models reviewed and suggested retail prices:

P212A	$595.00
P212C	$650.00
Star 103	$199.00

P212A

BAUD: 1200/300
COMPUTER: Any RS-232C-compatible system
TYPE: Bell 212A
SOFTWARE: Not included
AUTO ANSWER: Yes
AUTO DIAL: No
PHONE: RJ45S, RJ41S, (programmed), RJ21S (fixed loop), or RJ11C (permissive)

Features. The Tek-Com P212A modem provides full-duplex transfer of data at 0-300 baud using frequency-shift keying, and 1200 baud using phase-shift keying. The 1200-baud data transfer can be in either asynchronous or synchronous mode.

Three types of connections to the phone lines are possible: programmed, fixed-loss loop, and permissive. Tek-Com's programmed Vf cable uses a programming resistor, in either a Telco programmed jack (USOC RJ45S single or multi-jack) or a universal jack (USOC RJ41S single or multi-jack). In a fixed-loss loop arrangement, the Tek-Com fixed-loss loop Vf cable connects to a telco universal jack (USOC RJ21S) switched to the fixed-loss loop configuration. In the permissive arrangement the cable connects directly to the standard RJ11C modular connector.

Using the modem in the originate mode requires an exclusion-key telephone to switch between data and voice modes.

Discussion. A series of internal DIP switches configures the modem for specific applications. The following features are implemented through the DIP switches:

- Answer or originate mode.
- Interface speed indication.
- Transmit-clock available.
- Recieve-clock available.
- Carrier-detect and clear-to-send independence.
- Busy-out for analog loopback test.
- Data-terminal-ready forced on.
- Data-set-ready on for local analog loopback test.
- Internal or external transmit-clock source.
- Disconnect 350 milliseconds after loss of carrier frequency.
- Disconnect on loss of data terminal ready.
- 9- or 10-bit character length (including start and stop bits).
- Answer-mode indication.
- Auto-answer.
- Response to remote digital loopback test.
- Disconnect on 1.5-second incoming space.
- Activation of remote digital loopback test.
- Interface speed control.
- Activation of local analog test.

A series of external switches and LEDs can be used to implement local digital and remote digital loopback tests, local analog loopback tests, and self-tests. LEDs indicate carrier detect, speed, receiving data, transmitting data, data set ready, and data terminal ready.

P212C

BAUD: 1200/300
COMPUTER: Any RS-232C-compatible system
TYPE: Bell 212A
SOFTWARE: Not included
AUTO ANSWER: Yes
AUTO DIAL: No
PHONE: RJ45S, RJ41S (programmed), RJ21S (fixed loop), or RJ11C (permissive)

Features. The Tek-Com P212A modem provides full-duplex transfer of data at 0-300 baud using frequency-shift keying, and 1200 baud using phase-shift keying. The 1200-baud data transfer can be in asynchronous or synchronous mode.

The Tek-Com P212C has all of the features of the P212A including microprocessor control, auto self-test in idle mode, delayed busy-out and auto busy-out, remote alarm indicator, and manual speed override of auto-speed selection in answer mode.

The P212C will delay the analog loopback test and busy-out until the end of the current connection when the modem is off the hook. The P212C tests itself whenever it is in idle mode.

Three types of connections to the phone lines are possible: programmed, fixed-loss loop, and permissive. Tek-Com's programmed Vf cable uses a programming resistor in either a telco-programmed jack (USOC RJ45S single or multi-jack) or a universal jack (USOC RJ41S single or multi-jack). In a fixed-loss loop arrangement, the Tek-Com fixed-loss loop Vf cable connects to a telco universal jack (USOC RJ21S) switched to the fixed-loss loop configuration. In the permissive arrangement, the cable connects directly to the standard RJ11C modular connector.

To use the modem in the originate mode, you need an exclusion key telephone so you can switch between data and voice modes.

Discussion. A series of internal DIP switches configures the modem for specific applications. The following features are selected through the DIP switches:

- Answer or originate mode.
- Interface speed indication.
- Transmit-clock available.
- Receive-clock available.
- Carrier-detect and clear-to-send independence.
- Busy-out for analog loopback test.
- Data-terminal-ready forced on.
- Data set ready on for local analog loopback test.
- Internal or external transmit-clock source.

• Disconnect 350 milliseconds after loss of carrier frequency.
• 9- or 10-bit character length (including start and stop bits).
• Answer-mode indication.
• Auto-answer.
• Response to remote digital loopback test.
• Disconnect on 1.5 second incoming space.
• Disconnect 600 milliseconds after line-current interruption.
• Activation of remote digital loopback test.
• Interface speed control.
• Activation of local analog test.

A series of external switches and LEDs can be used to implement local digital and remote digital loopback tests, local analog loopback tests and self-tests. LEDs indicate carrier detect, speed, receiving data, transmitting data, data set ready, and data terminal ready.

Star 103

BAUD: 300
TYPE: Bell 103, CCITT V.21
COMPUTER: Any RS-232C-compatible system
SOFTWARE: Not included
AUTO ANSWER: No
AUTO DIAL: No
PHONE: Acoustically coupled

Features. This Bell 103-compatible modem acoustically couples to the handset of a standard telephone. The modem can transfer data at rates of 0-300 baud.

A three-position switch designates the Star 103 modem for answer mode, originate mode, or off. A second switch selects between full-duplex, half-duplex, and test mode.

Four LEDs indicate the status of the modem:

• The modem is transmitting.
• Data is being received.
• The remote modem is transmitting a proper carrier frequency and the Star 103 is ready for data transmission.
• The modem is currently being tested.

Discussion. Installation of the Star 103 requires the additional purchase of an RS-232C cable to connect the modem with a compatible computer.

With the Star 103 properly connected to a terminal or computer, a call is placed by dialing the appropriate number on the telephone. When the carrier frequency is heard, the handset of the phone is placed in the Star 103's acoustic cups. With the appropriate software running on your computer, data can now be exchanged between the remote and local computers.

A series of test procedures, including an analog loopback test, can be used to verify the working condition of the modem or to isolate problems with the modem, interface, or computer.

TELEVIDEO SYSTEMS, INC.

1170 Morse Ave.
Sunnyvale, CA 94086
(408) 745-7760

Televideo is one of the more successful manufacturers of terminals and terminal equipment. Its 1200-baud modem is an option for its Televideo Personal Terminal. Products are warranted against defects in parts and workmanship for a period of 90 days.

Model reviewed and suggested retail price:
Personal Terminal 300/1200-baud $549.00

Personal Terminal 300/1200 Modem

BAUD: 300/1200
COMPUTER: Televideo Personal Terminal
TYPE: Bell 212A
SOFTWARE: Included
AUTO ANSWER: Yes
AUTO DIAL: Yes (pulse- or tone-dial)
PHONE: RJ11 (modular) connection

Features. The Televideo modem is sold as an

option for the Televideo Personal Terminal. The modem operates at a selectable 300-baud or 1200-baud rate. When included, it provides the terminal with auto-answer and auto-dial capabilities. Dial and answer modes are entered by pressing special function keys on the terminal.

Dialing commands are entered as single characters from the keyboard and include a redial command that will automatically dial the last number again. The connected phone can be used either for data or voice communications.

Discussion. Dialing commands, entered from the keyboard, include: "T" for touch-tone dialing; "P" for pulse-tone dialing; "." for a two-second delay, to wait for a second dial tone, and "A/" to redial a previous number.

Additional keyboard commands allow you to program various timing variables: how many times the Personal Terminal's phone will ring before automatically answering, how long it will wait for a carrier tone when making a call, etc.

The status of the calls can be displayed on the screen in words or numbers, or not at all. After placing a call manually from the phone, you can transfer control to the modem to transfer data.

THE MICROPERIPHERAL CORPORATION

2565 152d Ave. N.E.
Redmond, WA 98052
(206) 881-7544

The Microperipheral Corp., in addition to the Microconnection, makes cables and acoustic couplers for its modems. The Microconnection is warranted against defects in materials and workmanship, except for normal wear and tear, for the lifetime of the modem.

Model reviewed and suggested retail price:
The Microconnection $179.95

The Microconnection

BAUD: 300
COMPUTER: ATARI, Commodore 64, VIC 20
TYPE: Bell 103
SOFTWARE: Includes program listing
AUTO ANSWER: Yes
AUTO DIAL: Yes (tone-dial)
PHONE: RJ11 (modular) connection

Features. This 300-baud modem is designed for use with Commodore 64, VIC 20, and ATARI computers (Fig. 8-40). It is Bell 103 compatible and has auto-dial, auto-answer, originate, and answer modes. This modem provides a parallel Centronics output port, allowing information received through the modem to transmit directly to a properly interfaced printer. Transmitted information displays simultaneously on the screen.

The modem connects to a modular (RJ11)

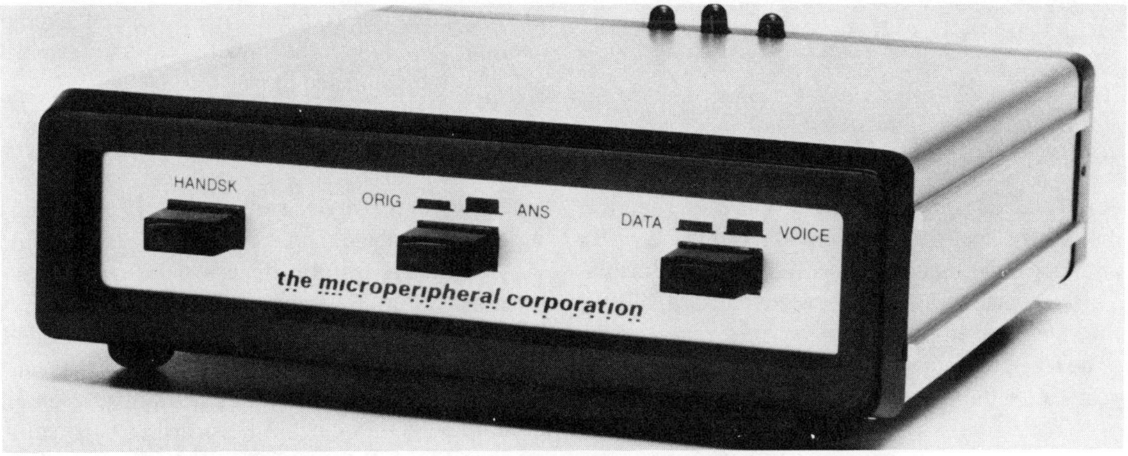

Fig. 8-40. The Microconnection modem. Courtesy The Microperipheral Corporation.

phone set. A single switch selects between using the phone for voice or data communications. LEDs indicate power-on, on-line status, and carrier-detect signals.

Discussion. The Microconnection includes cables for direct connection to the Commodore 64 through the user port; it does not require an additional RS-232C interface. The modem is powered through the computer and does not require a battery or a 120-volt ac outlet. It includes a cable for connection to an RJ11 phone socket; connecting a phone to the modem, however, requires the additional purchase of a T-connector.

When used with a printer, a special parallel cable, available through Radio Shack (part number 26-1409), must be purchased separately.

The documentation includes a BASIC program listing for utilizing the auto-answer, auto-dial, and print features of the modem. Data received through the modem can be sent directly to a printer capable of printing at greater than 300 baud. Some daisy-wheel printers are slower than 300 baud and therefore cannot be used for this feature.

The auto-answer feature of the modem allows you to leave the computer unattended and receive messages from other computers automatically. The auto-dial feature will automatically dial a phone number entered from the keyboard of your computer. The program listing must be manually typed into the computer and saved on whatever mass-storage device your computer has.

TNW CORPORATION

3444 Hancock St.
San Diego, CA 92110
(619) 296-2115

TNW is a company with a 6-year track record in the microcomputer communications industry. Its interfaces and modems are used by such Fortune 500 companies as IBM, Hewlett-Packard, TRW, Rockwell, and Dupont. TNW warrants its products to be free of manufacturing defects for a period of 2 years after the date of original purchase.

The model reviewed and suggested retail price:

Operator 103	$169.00

Operator 103

BAUD: 300
COMPUTER: Any RS-232C-compatible system
TYPE: Bell 103
SOFTWARE: Not included
AUTO ANSWER: Yes
AUTO DIAL: Yes (pulse-dial)
PHONE: RJ11 (modular) connection

Features. This Bell 103-compatible modem is designed for use with any RS-232C-compatible system. An "intelligent modem," it recognizes nine commands: answer, dial, return, hangup, baud, echo, fdial, intercept, and status.

It is capable of transmitting at rates of up to 300 baud. It has auto-dial, auto-answer, originate, and answer modes.

The modem connects directly to a modular (RJ11) phone set. A special Y adaptor can be purchased separately to provide a second connection for an existing phone.

Discussion. The modem operates in two modes: command and terminal. In the command mode, characters entered from the keyboard are sent to the modem, and characters which correspond to commands result in the execution of those commands. In the terminal mode, characters transmit through the modem to a receiving system. A special intercept character, designated by the operator, will switch the modem from the terminal mode to the command mode.

Seven DIP switches allow hardware configuration of the following parameters: baud rate 300, auto-answer disabled, echo disabled, normal dial speed, modem carrier-sense normal, command-interpreter enabled, and command-output enabled.

A fast dial feature enables pulse-dialing at a much faster rate than normal. An auto-answer feature allows you to leave the computer unattended and receive messages from other computers automatically. An auto-dial feature allows you to enter a phone number from the keyboard of your computer and have that number dialed automatically.

TRANSEND CORPORATION

2190 Paragon Dr.
San Jose, CA 95131
(408) 946-7400

Founded in 1974, Transend manufactures and markets a wide range of interface cards, electronic communications hardware, and software for the IBM PC and Apple computer. The company also manufactures and markets board-level products for the IEEE 696/S-100 and IEEE 488 bus.

Transend warrants its products to original purchasers to be free from defects in materials or workmanship for a period of 2 years. Products can be returned to the company for service or replacement.

Models reviewed and suggested retail prices:

PC Modemcard 1200	$549.00
Transmodem 1200	$695.00
Apple Modemcard	$275.00

PC Modemcard 1200

BAUD: 1200/300
COMPUTER: IBM PC, IBM XT
TYPE: Bell 212A, 103
SOFTWARE: Included
AUTO ANSWER: Yes
AUTO DIAL: Yes (pulse- or tone-dial)
PHONE: RJ11C (modular) connection

Features. The Modemcard is an internal modem that inserts into any of the full-size expansion slots of the IBM PC or XT (Fig. 8-41). It is Bell 103 compatible at 300 baud, and Bell 212A compatible at 1200 baud. It features auto-dial and auto-answer and recognizes Hayes modem commands. It automatically detects baud rate in originate or answer modes and adjusts speed accordingly. The modem includes two RJ11C jacks that allow you to leave a telephone connected for normal use when the computer is not on line.

The modem includes auto-dial features, using either pulse-dial or tone-dial systems. Numbers can be dialed from the keyboard, using any of the Hayes commands or through appropriate software.

The PC212A supports auto-answer features, implemented through software, and monitors the progress of calls with messages such as busy signal, ringing, on-line, etc.

This modem includes a terminal program and a subscription to *THE SOURCE*. The modem is compatible with Transcend PC and is sold with the software as Transpak PC for a total of $659.00.

Discussion. The documentation provides instructions for installing the modem into an available slot inside the computer. The modem can be set to occupy any 8-byte boundary within 1024 possible I/O addresses enabling you to alter the address if a serial card already occupies the configured address. Most communications software uses an input-

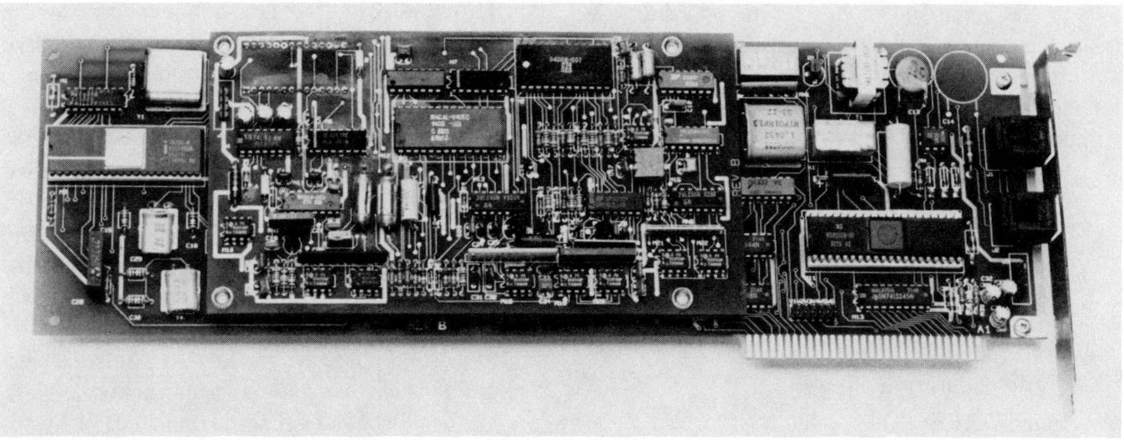

Fig. 8-41. The PC Modemcard 1200. Courtesy Transend Corporation.

interrupt buffer-storage scheme to reduce the possibility of missing an incoming character. The Modemcard has six jumpers for each of the interrupt-request lines in your personal computer for the modem.

The Modemcard has a complete set of commands for Hayes-compatible software. For example, the modem determines the baud rate when the command "A" is entered, and the parity and word length when the letter "T" is entered. All other commands are preceded by "AT." Commands concerning the auto-answer capabilities will answer the phone automatically, before it actually rings or after any given number of rings. Dialing commands from the keyboard can direct a call over pulse or touch-tone systems. Pauses can be implemented in the dialing sequence to wait for a second dial tone. Responses to the commands entered from the keyboard can be displayed on the screen, in either single-character or full-word format. Additional commands allow you to configure RS-232C parameters from the keyboard.

Registers which can be programmed from the keyboard control various aspects of the modem: the number of rings before auto-answer, the time to wait for a dial tone, the time after completing a call to wait for a valid carrier, and the time after the loss of a carrier tone when the modem automatically hangs up.

The modem operates in two states: the command mode and the on-line mode. The modem assumes the command mode, for recognizing keyboard commands, when it is powered up. It will automatically switch to the on-line mode, for transferring data, after a connection has been made with a compatible system.

Transmodem 1200

BAUD: 1200/300
COMPUTER: Any RS-232C-compatible system
TYPE: Bell 212A/103
SOFTWARE: Not included
AUTO ANSWER: Yes
AUTO DIAL: Yes (pulse- or tone-dial)
PHONE: RJ11C (modular) connection

Features. The Transmodem 1200 is an external modem that connects to most computers through standard RS-232C-type interfaces. The modem includes two RJ11C jacks that allow you to leave a telephone connected for normal use when the computer is not on line.

All activities are controlled through keyboard commands. The modem has no panel switches. A series of panel lights help monitor the progress of communications. The Transmodem 1200 will automatically sense the working baud rate and configure itself appropriately.

The modem includes auto-dial features, using either pulse-dialing or tone-dialing. Numbers can be dialed and redialed automatically. A number consisting of up to 32 digits can be loaded into the modem's memory and redialed at a later time. The Transmodem supports auto-answer features, implemented through keyboard commands, and monitors the progress of calls with messages such as DIALING, RINGING, VOICE, BUSY, FAILED CALL, etc.

This modem is compatible with Transend 1, 2, 3 and Transend PC. The modem is sold as a package with either of these communications programs at an increased price.

The modem can institute a local analog loopback test to verify its working condition.

Discussion. The Transmodem provides an external RS-232C port, for connecting a standard DB25 cable, and two RJ11C ports, for connection to the phone system.

The modem can be operated in four basic modes. The modem assumes an idle mode, where it can answer the telephone manually or automatically, in a normal start-up. An interactive mode allows the user to enter dialing commands, change software options, or change a number in memory. A data mode allows the transfer of data to or from the modem. A test mode allows the execution of a local analog loopback test. The various modes are entered either through keyboard commands or by automatic operation.

Commands for auto-dial provide for touch-tone dialing, pulse-tone dialing, pausing between numbers, and redialing. Numerous options are determined through software control. Baud rate, timing, and Bell compatibility are set through option commands. Originate or answer modes, as well as word length, are selected through software. The modem can be set to disconnect 1 second after the loss of the received-carrier signal or 1½ seconds after receiving continuous spaces.

Normally, a number is not dialed until the modem senses a dial tone. The Transmodem has the ability to blind dial for phones with nonstandard tones. Panel lights indicate when the modem is operating 1200-bauds, receiving data, transmitting data, or detecting a valid carrier.

Apple Modemcard

BAUD: 110/300
COMPUTER: Apple II, *II*e
TYPE: Bell 103
SOFTWARE: Not included
AUTO ANSWER: Yes
AUTO DIAL: Yes (pulse- or tone-dial)
PHONE: RJ11C (modular) connection

Features. The Apple Modemcard is an internal modem that inserts into any of the expansion slots of the Apple II, except slot 0 (Fig. 8-42). It is Bell 103 compatible and capable of full- or half-duplex transmissions at 110 or 300 baud. It features auto-dial and auto-answer, and includes an internal speaker for audio monitoring of the telephone. The modem includes one RJ11C jack, for direct connection to an appropriate wall outlet.

The auto-dial features use either pulse-dial or tone-dial systems. Firmware prompts for the number to be dialed, including any necessary delays. The internal speaker allows you to hear dial tones, busy signals, etc.

The Modemcard includes firmware for an auto-answer mode. The modem will automatically answer a call and allow the remote computer full access to your Apple computer.

This modem includes a subscription to *The SOURCE*. It is compatible with Transend 1, 2, and 3 communications software. For a slightly higher price, the modem is sold with that software as Transpak 1, 2, or 3. (See Chapter 9.)

Discussion. The Apple Modemcard has three modes of operation: terminal mode, remote mode, and programming mode. In the terminal mode, various options can be set, including baud rate and

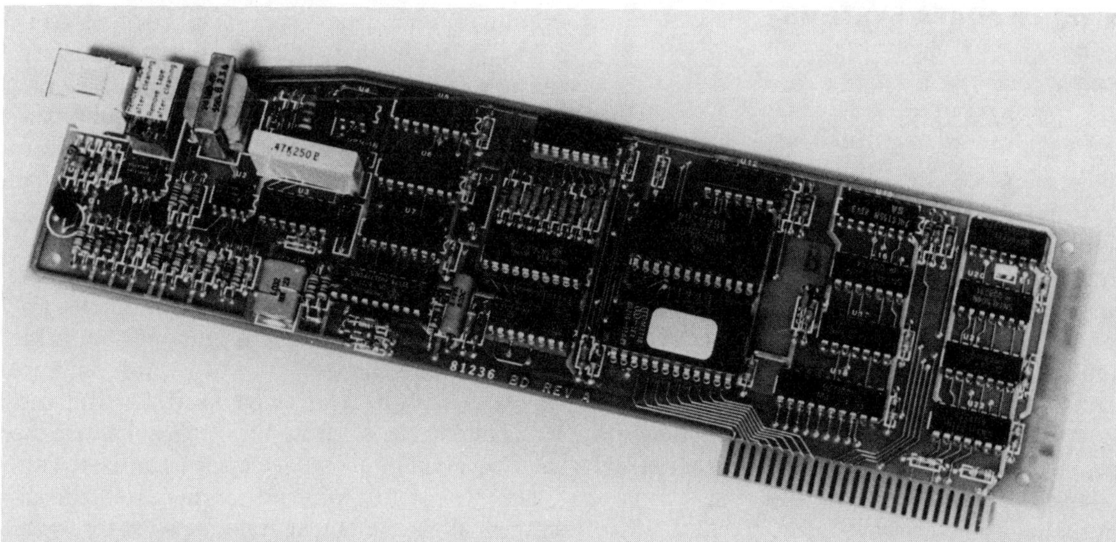

Fig. 8-42. The Apple Modemcard. Courtesy Transend Corporation.

duplex. If desired, entries from the keyboard can be directed to a printer, as well as to the screen. Options are selected through simple keyboard commands. Within the terminal mode, pressing "Ctrl A Ctrl Q" results in a prompt: ENTER PHONE NUMBER.

Numbers are dialed automatically, according to additional commands: "T" causes touch-tone dialing; "P" causes pulse-dialing; "A" activates the audio monitor, and "*" inserts a 2-second delay. The progress of a call displays on the screen with messages such as AWAITING CARRIER, CONNECTION ESTABLISHED, NO CARRIER DETECTED, and PHONE HUNG UP.

Entering a series of special characters places the modem in the remote mode. The modem and computer can be left unattended to answer incoming calls. The status of incoming calls displays on the screen with messages such as LINE RINGING, AWAITING CARRIER, and CONNECTION ESTABLISHED. Once a connection is established, the remote caller has complete control of your computer. The caller can load, run, and list programs from our system. If the data is arriving too fast, the remote computer can temporarily halt transmission. If desired, the remote computer can reset the computer or put it into terminal mode.

UNIVERSAL DATA SYSTEMS

5000 Bradford Dr.
Huntsville, AL 35805
(205) 837-8100

Universal Data Systems, a unit of Motorola Corporation Information Systems Group, manufactures over 20 different modems. Since 1970, the company has been providing communications equipment to consumers and original equipment manufacturers. It claims that less than ½ of 1 percent of its modems are returned for repair. The company warrants, to the original purchaser, its equipment to be free from defects in material and workmanship for 1 year from date of sale. It provides a toll-free technical-support number: 1-800-221-4380.

Models reviewed and suggested retail prices.

108	$295.00
103J	$425.00
202S	$475.00
202S/SS	$550.00
212A	$595.00
212A/D	$645.00
212LP	$345.00

108

BAUD: 300
COMPUTER: Any RS-232C-compatible system
TYPE: Bell 103
SOFTWARE: Not included
AUTO ANSWER: No
AUTO DIAL: No
PHONE: RJ11C (modular) connection private line only

Features. The Universal Data Systems 108 has frequency-shift keying for asynchronous operation at rates up to 300 baud. It operates for half-duplex or full-duplex through private line connections. It comes in a stand-alone case, or as a single board for rack-mount via an adapter.

The front-panel display includes five LEDs to indicate transmit data, receive data, carrier detect, power on, and error detect. A rotary switch selects among a self-test, analog loopback test, digital loopback test, digital loopback test, testing patterns, and data transmission.

The rear panel includes connections for power, an RS-232C connection, an RJ11C connection, and a fuse. The modem can be configured for various modes of operation through strap settings located inside it. The documentation includes instructions on how to access these settings and make the appropriate changes.

Discussion. The Universal Data Systems 108 is intended for private-line applications. Internal strap settings allow the device to be used with either two- or four-wire private lines. Additional settings allow the output transmit level to be altered to either 0, −3, −6, −9, and −12 dbm. The signal ground can be connected to the chassis ground through an internal strap. Answer and originate modes are also determined through internal switch settngs.

Additional configurations allows two choices of clear-to-send delay timing and carrier-detect delay timing. The transmitter can be placed under terminal control or can be turned on continuously. The modem can be configured for various line impedance and used in multi-drop connections.

A self-test mode generates a test pattern inside the modem and verifies accuracy. The test does not transmit data onto the line or through the RS-232C interface; it confines the test to the modem itself.

An analog loopback test loops characters from the keyboard through the modem and back to the display for verification of the modem, interface, and computer. A digital loopback test loops data from the phone lines back onto the line without passing through the RS-232C interface. This test is used in cooperation with the operator at the other end.

103J

BAUD: 300
COMPUTER: Any RS-232C-compatible system
TYPE: Bell 103/113
SOFTWARE: Not included
AUTO ANSWER: Yes
AUTO DIAL: No
PHONE: RJ11C (modular) connection or others by request

Features. The Universal Data Systems 103J has frequency-shift keying for asynchronous operation at rates of up to 300 baud. It is designed for half-duplex or full-duplex operation through the public phone system or private-line connections (optional). It comes in a stand-alone case or as a single board for rack-mount via an adapter.

The front panel displays five LEDs for indicating transmit data, receive data, carrier detect, power on, and ringing current present. A rotary switch selects between a local and remote test position, and data and voice transmission.

The rear panel includes connections for power, an RS-232C connection, an RJ11C telco connection to an appropriate wall jack, an RJ11C tel-set connection to a compatible phone, and a fuse. The modem can be configured for various modes of operation through strap settings located inside it. The documentation includes instructions on how to access these settings and make the appropriate changes.

Discussion. The Universal Data System 103J is intended to be used with the public phone system, but it can be used over private lines with appropriate modifications.

Internal strap settings allow the device to be used with either two-wire switched networks or four-wire private lines. Additional settings allow the output-transmit level to be altered between −9 dbm and a programmable level dependent on modifications. The signal ground can be connected to the chassis ground through an internal strap. Answer and originate modes are also determined through internal switch settings. Other configurations allow the call to be disconnected automatically if the carrier signal is lost.

The front-panel switch allows the user to select between the data mode and the talk mode. In the talk mode, the phone is essentially removed from the modem circuit and used normally for voice communication. In the data mode, the phone is used in conjunction with the modem for data transmission.

Front-panel LEDs indicate various conditions:

- Detecting a carrier frequency from another modem.
- Receiving data and passing it on to the terminal.
- Receiving data from the terminal.
- Ringing current present.

A local test, implemented by a front-panel switch, loops characters from the keyboard through the modem and back to the display, for verification of the modem, interface, and computer. A remote test, also implemented from the front panel, loops data from phone lines back onto the line, without passing through the RS-232C interface. This test is used in cooperation with the operator at the other end.

202S

BAUD: 0-1200
COMPUTER: Any RS-232C-compatible system
TYPE: Bell 202
SOFTWARE: Not included
AUTO ANSWER: Yes
AUTO DIAL: No
PHONE: RJ11C (modular) connection, or others by request

Features. The Universal Data Systems 202S has frequency-shift keying for asynchronous operation at 0 to 1200 baud. It is intended for half-duplex operation through the two-wire public phone system, or for full-duplex operation through four-wire private-line connections. It comes in a stand-alone case or as a single-board for rack-mount via an adapter.

The front panel displays five LEDs that indicate transmit data, receive data, carrier detect, power on, and off hook. A rotary switch selects among a self-test, analog loopback test, and remote test, and data and voice transmission. The rear panel includes connections for power, an RS-232C connection, an RJ11C telco connection to an appropriate wall jack, an RJ11C tel-set connection to a compatible phone, and a fuse.

The modem can be configured for various modes of operation through strap settings located inside the modem. The documentation includes instructions on how to access these settings and make the appropriate changes.

Discussion. The Universal Data Systems 202S is intended for use on the public phone system, but it can be used on private lines as well. Internal strap settings allow the device to be used with either two-wire switched networks or four-wire private lines. Additional settings allow the output transmit level to be altered between −9 dbm and a programmable level dependent on modifications. The signal ground can be connected to the chassis ground through an internal strap.

Additional configurations allow a call to be disconnected automatically if the carrier is lost. Also, the modem can be used to answer a call automatically or manually.

The front-panel switch allows the user to select between the data mode and the talk mode. In the talk mode, the phone is used normally for voice communication. In the data mode, the phone is used in conjunction with the modem for data transmission.

Front-panel LEDs indicate various conditions:

- Detecting a carrier frequency from another modem.
- Receiving data and passing it on to the terminal.
- Receiving data from the terminal.
- Ringing current present.

A local test, implemented from a switch on the front panel, loops characters from the keyboard through the modem and back to the display, to verify the modem, interface, and computer.

With the remote test on, the RS-232C interface disconnects, the data-terminal-ready signal goes high, and the modem waits for a call. In receiving the ring indication, the modem goes off-hook, delays for a couple seconds, and then sends a test pattern to the calling party. This test is used in cooperation with the operator at the other end.

202S/SS

BAUD: 0-1200
COMPUTER: Any RS-232C-compatible system
TYPE: Bell 202
SOFTWARE: Not included
AUTO ANSWER: Yes
AUTO DIAL: No

PHONE: RJ11C (modular) connection, or others by request

Features. The Universal Data Systems 202S/SS has frequency-shift keying for asynchronous operation, at 0 to 1200 baud, or for synchronous operation at 600 baud or 1200 baud. It is intended for half-duplex operation through the two-wire public phone system, or for full-duplex operation through four-wire private line connections. It comes in a stand-alone case or as a single board for rack-mount via an adapter.

The front panel displays five LEDs that indicate transmit data, receive data, carrier detect, power on, and off hook. A rotary switch selects among a self-test, analog loopback test, and remote test, and data and voice transmission. The rear panel includes connections for power, an RS-232C connection, an RJ11C telco connection to an appropriate wall jack, an RJ11C tel-set connection to a compatible phone, and a fuse.

The modem can be configured for various modes of operation through strap settings located inside the modem. The documentation includes instructions on how to access these settings and make the appropriate changes.

Discussion. The Universal Data Systems 202S/SS is intended for use with the public phone system, but it can be used over private lines as well.

Internal strap settings allow the device to be used with either two-wire switched networks or four-wire private lines. Additional settings allow the output transmit level to be altered between −9 dbm and a programmable level dependent on modifications. The signal ground can be connected to the chassis ground through an internal strap.

Additional configurations allow a call to be disconnected automatically if the carrier is lost. The modem can be used to answer a call automatically or manually.

A switch on the front panel allows the user to select between the data mode and the talk mode. In the talk mode, the phone is used normally for voice communication. In the data mode, the phone is used in conjunction with the modem for data transmission.

Front-panel LEDs indicate various conditions:

- Detecting a carrier frequency from another modem.
- Receiving data and passing it on to the terminal.
- Receiving data from the terminal.
- Ringing current present.

A local test, implemented from a switch on the front panel, loops characters from the keyboard through the modem and back to the display, verifying the modem, interface, and computer. With the remote test on, the RS-232C interface disconnects, the data-terminal-ready signal goes high, and the modem waits for a call. On receiving the ring indication, the modem goes off-hook, delays for a couple seconds, and then sends a test pattern to the calling party. This test is used in cooperation with the operator at the other end.

212A

BAUD: 0-300, 1200
COMPUTER: Any RS-232C-compatible system
TYPE: Bell 103/113/212
SOFTWARE: Not included
AUTO ANSWER: Yes
AUTO DIAL: No
PHONE: RJ11C (modular) connection or others by request

Features. The Universal Data Systems 212A provides full-duplex operation on two-wire switched networks or two-wire private lines. It operates at 0-300 baud asynchronously, using a Bell 103 or 113 format, or at 1200 baud, synchronously or asynchronously, using a Bell 212 format. It comes in a stand-alone case or as a single board for rack mounting using an optional adapter.

The front panel displays eight LEDs that indicate the operating status of the modem: make busy, terminal ready, modem ready, transmit data, receive data, high/low speed, modem check, and test mode. Six buttons select various options: a self-test, analog loopback test, remote digital loopback test, local digital loopback test, high speed and remote test, and data transmission. The rear panel includes connections for power, an RS-232C connection, an RJ11C telco connection to an appropriate wall jack, an RJ11C tel-set connection to a compatible phone, and a fuse.

The modem can be configured for various modes of operation through strap settings located inside the modem. The documentation includes instructions on how to access these settings and make the appropriate changes.

Discussion. The Universal Data Systems 212A uses frequency-shift keying at 0-300 baud, or differential phase-shift keying at 1200 baud. Internal switch settings designate its use for public phone system or private lines.

Normally, the maximum allowable output of a modem over public lines is −9dbm. Losses in output because of line impedance to the nearest central phone station can be compensated with "programming" resistors. The 212A has internal strap settings to select between the −9dbm and a programmable state.

In synchronous operation, either external or internal clocks can be selected through strap settings, and a signal ground can be connected to the chassis ground through a separate strap. Additional settings select other options; answering the phone manually or automatically; 9-, 10-, or 11-bit format; synchronus or asynchronous operation; and public line or private line.

Buttons on the front panel allow the user to select between the data mode and the talk mode. In the talk mode, the phone is used normally for voice communication. In the data mode, the phone is used in conjunction with the modem for data transmission.

Front-panel LEDs indicate various conditions:

- Receiving data and passing it to the terminal equipment.
- Receiving data from the terminal.
- Analog loopback mode.
- Terminal active.
- Modem ready to transmit or receive data.
- Operating at 1200 baud.
- Errors during a self-test.
- The test mode is on.

The analog loopback test isolates the modem from the telephone lines and loops the transmit signal back through the receive circuitry for display on your video display unit. The self-test can be used with the digital tests to transmit a test pattern and to check received data for errors.

The remote digital loopback signals the modem at the other end of the line to enter the digital loopback mode. Characters transmitted from your terminal will be looped back from the distant modem for verification.

The digital loopback takes data received over the phone lines and loops it back over the lines to the originating system.

212A/D

BAUD: 0-300, 1200
COMPUTER: Any RS-232C-compatible system
TYPE: Bell 103/113/212
SOFTWARE: Not included
AUTO ANSWER: Yes
AUTO DIAL: Yes (pulse- or tone-dial)
PHONE: RJ11C (modular) connection or others by request

Features. The Universal Data System 212A-/D provides full-duplex operation on two-wire switched networks or two-wire private lines. It operates at 0-300 baud asynchronously, using a Bell 103 or 113 format, or at 1200 baud, synchronously or asynchronously, in a Bell 212 format. It can dial from the keyboard of the terminal or from five numbers stored in the modem's memory. The numbers can include up to 30 digits or commands. The 212A/D comes in a stand-alone case or as a single board for rack mounting with an optional adapter.

The front panel displays eight LEDs that indicate the operating status of the modem: make busy, terminal ready, modem ready, transmit data, receive data, high/low speed, modem check, and test mode. Six buttons select various options: a self-test, analog loopback test, remote digital loopback test, logical digital loopback test, high speed and remote test, and data transmission. The rear panel includes connections for power, an RS-232C connection, an RJ11C telco connection to an appropriate wall jack, an RJ11C tel-set connection to a compatible phone, and a fuse.

The modem can be configured for various modes of operation through strap settings located inside the modem. The documentation includes instructions on how to access these settings and make the appropriate changes.

Discussion. The Universal Data System 212A/D uses frequency-shift keying at 0-300 baud, or differential phase-shift keying at 1200 baud. Internal switch settings designate its use for the public phone system or private lines.

The 212A/D recognizes 11 keyboard commands for various dialing features. The commands include "D" for dial, "W" for wait for dial tone, "L" for load numbers into memory, "C" for clear memory, "R" for repeat last number dialed, "X" for repeat dialing until answered, "P" for apuse, and "H" for display all commands. These commands can come directly from the keyboard, or they can be stored into memory along with phone numbers. A single keyboard command can then issue the commands to the modem along with one of the five stored numbers.

Normally, the maximum allowable output of a modem over public lines is −9dbm. Losses in output because of line impedance to the nearest central phone station can be compensated with "programming" resistors. The 212A has internal strap settings to select between the −9dbm and a programmable state.

In synchronous operation, either external or internal clocks can be selected through strap settings. Through a separate strap, a signal ground can be connected to the chassis ground. Additional settings make other selections: answering the phone manually or automatically; 9-, 10-, or 11-bit format; synchronous or asynchronous operation; and public line or private line use.

Buttons on the front panel allow the user to select between the data mode and the talk mode. In the talk mode, the phone is used normally for voice communication. In the data mode, the phone is used in conjunction with the modem for data transmission.

Front-panel LEDs indicate various conditions:

- Receiving data and passing it to the terminal equipment.
- Receiving data from the terminal.
- Operating in the analog loopback mode.
- Data-terminal-ready line from the terminal is active.
- Modem ready to transmit or receive data.
- Operating at 1200 baud.
- Errors during a self-test.
- The test mode is on.

The analog loopback test isolates the modem from the telephone line and loops the transmit signal back through the receive circuitry for display on your video display unit. The self-test can be used with the digital tests to transmit a test pattern and to check received data for errors.

The remote digital loopback signals the modem at the other end of the line to enter the digital loopback mode. Characters transmitted from your terminal will be looped back from the distant modem for verification. The digital loopback takes data received over the phone lines and loops it back over the lines to the originating system.

212 LP

BAUD: 1200
COMPUTER: Any RS-232C-compatible system
TYPE: Bell 212 (1200 baud only)
SOFTWARE: Not included
AUTO ANSWER: No

AUTO DIAL: No
PHONE: RJ11C (modular) connection

Features. The Universal Data Systems 212 LP provides full-duplex 1200-baud operation on two-wire switched networks. It is powered through voltages received from the telephone line and requires no additional power supply.

The rear panel accommodates an RS-232C connection, an RJ11C telco connection to an appropriate wall jack, and an RJ11C tel-set connection to a compatible phone. The modem operates in originate or answer modes and data or talk modes.

Discussion. The Universal Data System 212 LP operates in a full-duplex mode over public lines at a single speed of 1200 baud, using a differential-coherent phase-shift keying. It is compatible with Bell 212A operating in an asynchronous mode. Word length is programmable, through internal switch settings, to 9, 10, or 11 bits, including start and stop bits.

The modem can be set manually, from convenient switches on the front panel, for originate or answer modes and data or talk modes. In the talk mode, the phone is used normally for voice communication. In the data mode, the phone is used in conjunction with the modem for data transmission.

U.S. ROBOTICS, INC.

1123 W. Washington Blvd.
Chicago, IL 60607
(312) 733-0497

U.S. Robotics, founded in 1976, designs, develops, manufactures, and markets modems. The company's product line currently includes nine modems and one communications program.

U.S. Robotics provides a customer-support number (1-312-733-1053) to provide assistance for customers experiencing problems with a modem. If necessary, products can be returned for repair with a one-day turn-around time.

Models reviewed and suggested retail prices:

Password	$449.00
Password 300	$199.00
Personal Communicator	$499.00
S-100	$449.00

Password

BAUD: 1200/300
COMPUTER: Any RS-232C-compatible system
TYPE: Bell 212A/103
SOFTWARE: Not included
AUTO ANSWER: Yes
AUTO DIAL: Yes (pulse- or tone-dial)
PHONE: RJ11C

Features. Password is an external 1200-baud or 300-baud Bell 212A/103-compatible modem that will interface with almost any computer with RS-232C capabilities. Password will automatically dial and answer calls. It is compatible with most communications software, including TELPAC (also from U.S. Robotics) and XMODEM (in the public domain).

The modem has a single RJ11C port for connection to an appropriate wall outlet. A duplexer can be purchased and connected to the wall outlet for simultaneous connection to a phone.

Four DIP switches, located in the rear, make special configurations:

- Keeping the DTR line open.
- Using a PBX system.
- Using a private line.
- Using a terminal, rather than a computer.

Discussion. Dialing commands enable pulse- or tone-dialing. A sequence of commands, up to 60 characters long, can be transferred to the modem at one time. The progress of a call (busy tones, etc.) can be monitored through an internal speaker. The speaker's loudness can be controlled through special commands.

In addition to audible monitoring, short messages display to indicate the progress of calls and the results of various commands: OK (when a command has been executed); CONNECT (when a connection to another system has been made); etc.

An answer mode will automatically answer a call after a preset number of rings. If the incoming carrier frequency remains absent after 17 seconds, the modem will disconnect. The answer mode can be entered through a special command or through a DIP switch setting on the back of the modem.

Password 300

BAUD: 300
COMPUTER: Any RS-232C-compatible system
TYPE: Bell 103
SOFTWARE: Not included
AUTO ANSWER: Yes
AUTO DIAL: Yes (pulse- or tone-dial)
PHONE: RJ11C

Features. Password 300 is an external 300-baud Bell 103-compatible modem that will interface with almost any computer with RS-232C capabilities. Password will automatically dial and answer calls. It is compatible with most communications software, including TELPAC (also from U.S. Robotics) and XMODEM (in the public domain).

The modem has a single RJ11C port for connection to an appropriate wall outlet. A duplexer can be purchased and connected to the wall outlet for simultaneous connection to a phone.

Four DIP switches, located in the rear, make special configurations:

- Keeping the DTR line on.
- Using a PBX system.
- Using a private line.
- Using a terminal, rather than a computer.

Discussion. Dialing commands enable pulse- or tone-dialing. A series of commands, up to 60 characters long, can be transferred to the modem at one time. The progress of a call (busy tones, etc.) can be monitored through an internal speaker. The speaker's loudness can be controlled through special commands.

In addition to audible monitoring, short messages are displayed indicating the progress of calls and the results of various commands: OK (when a command has been executed); CONNECT (when a connection to another system has been made); etc.

An answer mode will automatically answer a call after a preset number of rings. If the incoming carrier frequency remains absent after 17 seconds, the modem will disconnect. The answer mode can be entered through a special command or through a DIP switch setting on the back of the modem.

The Personal Communicator Modem

BAUD: 1200/300
COMPUTER: IBM PC, XT, or compatibles
TYPE: Bell 212A/103
SOFTWARE: Included (Telpac)
AUTO ANSWER: Yes
AUTO DIAL: Yes (pulse- or tone-dial)
PHONE: RJ11C

Features. The PC Modem is an internal 1200-baud or 300-baud Bell 212A/103-compatible modem (Fig. 8-43). It inserts into any of the available slots inside the IBM PC or XT. The modem includes Telpac communications software, but it works with most communications software for the IBM. It recognizes commands for auto-dial and auto-answer and can operate on both touch- and pulse-tone systems. The modem has two RJ11C ports for connection to an appropriate wall outlet and telephone.

Four DIP switches, located in the rear, make special configurations:

- Keeping the DTR line on.
- Using a PBX system.
- Using a private line.
- Using a terminal, rather than a computer.

Discussion. Dialing commands enable pulse- or tone-dialing. A sequence of commands, up to 60 characters long, can be transferred to the modem at one time. The progress of a call (busy tones, etc.) can be monitored through an internal speaker. The speaker's loudness can be controlled through special commands.

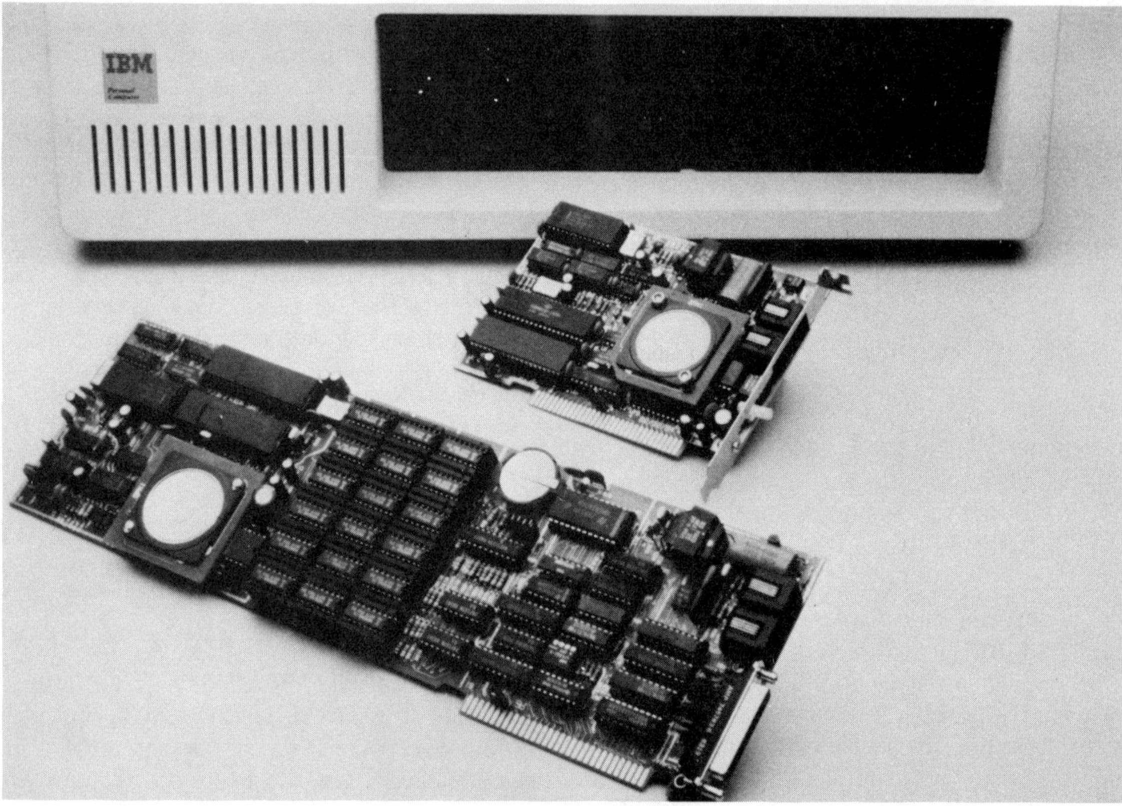

Fig. 8-43. The Personal Communication modem. Courtesy U.S. Robotics, Inc.

In addition to audible monitoring, short messages display to indicate the progress of calls and the results of various commands. An answer mode will automatically answer a call after a preset number of rings. If the incoming carrier frequency remains absent after 17 seconds, the modem will disconnect. The answer mode can be entered through a special command or through a DIP switch setting on the back of the modem.

S-100

BAUD: 1200/300
COMPUTER: S-100 computers
TYPE: Bell 212A/103
SOFTWARE: Not included
AUTO ANSWER: Yes
AUTO DIAL: Yes (pulse- or tone-dial)
PHONE: RJ11C

Features. The S-100 is an internal 1200-baud or 300-baud Bell 212A/103-compatible modem. It inserts into any computer using an S-100 bus. The modem is compatible with Telpac communications sofware.

It recognizes commands for auto-dial and auto-answer and can operate on both pulse- and touch-tone systems. The modem has two RJ11C ports for connection to an appropriate wall outlet and telephone.

Four DIP switches, located in the rear, make special configurations:

- Keeping the DTR line on.
- Using a PBX system.
- Using a private line.
- Using a terminal, rather than a computer.

Discussion. Dialing commands enable pulse-

or tone-dialing. A series of commands, up to 60 characters long, can be transferred to the modem at one time. The progress of a call (busy tones, etc.) can be monitored through an internal speaker. The speaker's loudness can be controlled through special commands.

In addition to audible monitoring, short messages display to indicate the progress of calls and the results of various commands. An answer mode will automatically answer a call after a preset number of rings. If the incoming carrier frequency remains absent after 17 seconds, the modem will disconnect. This mode can be entered through a special command or through a DIP switch setting on the back of the modem.

VEN-TEL, INC.

2342 Walsh Ave.
Santa Clara, CA 95051
(408) 727-5721

Ven-Tel has been designing, manufacturing, and marketing modems for over nine years. The company's product line includes a complete line of low-speed modems (0-2400 baud) and a series of asynchronous and synchronous digital line drivers. These modems are available in both stand-alone and rack-mount configurations. Ven-Tel also supplies custom modems to original equipment manufacturers for building into terminals.

Products are backed with a 1-year warranty against defective parts and workmanship. The company offers, free of charge through a service number, technical advice concerning installation and operation of its products.

Models reviewed and suggested retail prices:

PC Modem Half-Card	$549.00
PC Modem 1200	$499.00
1200 Plus	$495.00
HP 150 Internal Modem	$425.00

PC Modem Half-Card

BAUD: 1200/300
COMPUTER: IBM PC, IBM XT, IBM portable, Compaq
TYPE: Bell 212A/103
SOFTWARE: Included (Crosstalk-XVI)
AUTO ANSWER: Yes
AUTO DIAL: Yes (pulse- or tone-dial)
PHONE: RJ11 (modular) connection

Features. The PC Modem Half-Card is an internal modem that inserts into the half-sized slot in the IBM PC or IBM XT, leaving full-size slots available for expansion (Fig. 8-44). A modified board is available for the IBM portable. The modem includes a single jack for connecting to the modular wall socket.

The modem includes auto-dial features, using either pulse- or tone-dial systems. Working with the software included, a file can be created that will automatically dial a number and initiate a logon sequence.

The modem also includes an auto-answer feature which allows callers to establish contact and leave messages on an unattended system. An answer program can be created that will request passwords and verify the response. Once accepted onto the system, the caller can look at the disk directory, leave a message, or upload, or download a file.

This modem recognizes the Hayes command set and works with numerous other communications software, including Smartcom II and PC Talk III.

Discussion. The documentation provides thorough instructions on installing the modem into the computer. A set of DIP switches on the card provides the option of automatically answering incoming calls. Many of the functions of the modem are actually the result of the software Crosstalk (see Chapter 9).

The modem itself is fully programmable. It assumes one of two states: the local command mode, where commands controlling the activities are transmitted to the modem; or the on-line mode, where data is actually transferred through the modem. Most of the commands are transmitted by the software automatically, in response to files you create in advance. A special command mode allows

Fig. 8-44. The PC Modem Half-Card. Courtesy Ven-Tel, Inc.

you to transmit some of these control characters directly from the keyboard.

Files can be created to execute the auto-dial feature. Additional commands necessary for logging on to a system can be included in the file, as well as specific RS-232C parameters for that number. A logon sequence can recognize responses from the other system and respond accordingly.

Commands for auto-dial provide for touch-tone dialing, pulse-dialing, pausing for two seconds, repeating dialing isntructions and returning to command mode. Commands for answering include answering after one ring, answering after five rings, and answering immediately.

Through Crosstalk, your terminal can emulate a number of other types of terminals so you can access specific computer systems. Emulated terminals include Televideo 910/920, IBM 3101 ASCII terminal, Adds Viewpoint, and DEC VT-100.

PC Modem 1200

BAUD: 1200/300
COMPUTER: IBM PC, IBM XT, IBM portable, Compaq
TYPE: Bell 212A/103
SOFTWARE: Included (Crosstalk XVI)
AUTO ANSWER: Yes

AUTO DIAL: Yes (pulse- or tone-dial)
PHONE: RJ11 (modular) connection

Features. The PC Modem 1200 is an internal modem that inserts into any of the full-size expansion slots in the IBM PC, XT, portable, or the Compaq system (Fig. 8-45). The modem includes two RJ11 jacks that allow you to leave a telephone connected for normal use when the computer is not on-line.

The modem includes auto-dial features, using either pulse- or tone-dial systems. Working with the software included, a file can be created that will automatically dial a number and initiate a logon sequence. A built-in speaker allows you to monitor the progress of calls.

The modem also includes an auto-answer feature which allows callers to establish contact and leave messages on an unattended system. An answer program can be created that will request passwords and verify the response. Once accepted onto the system, the caller can look at the disk directory, leave a message, or upload or download a file.

This modem recognizes the Hayes command set and works with numerous other communications

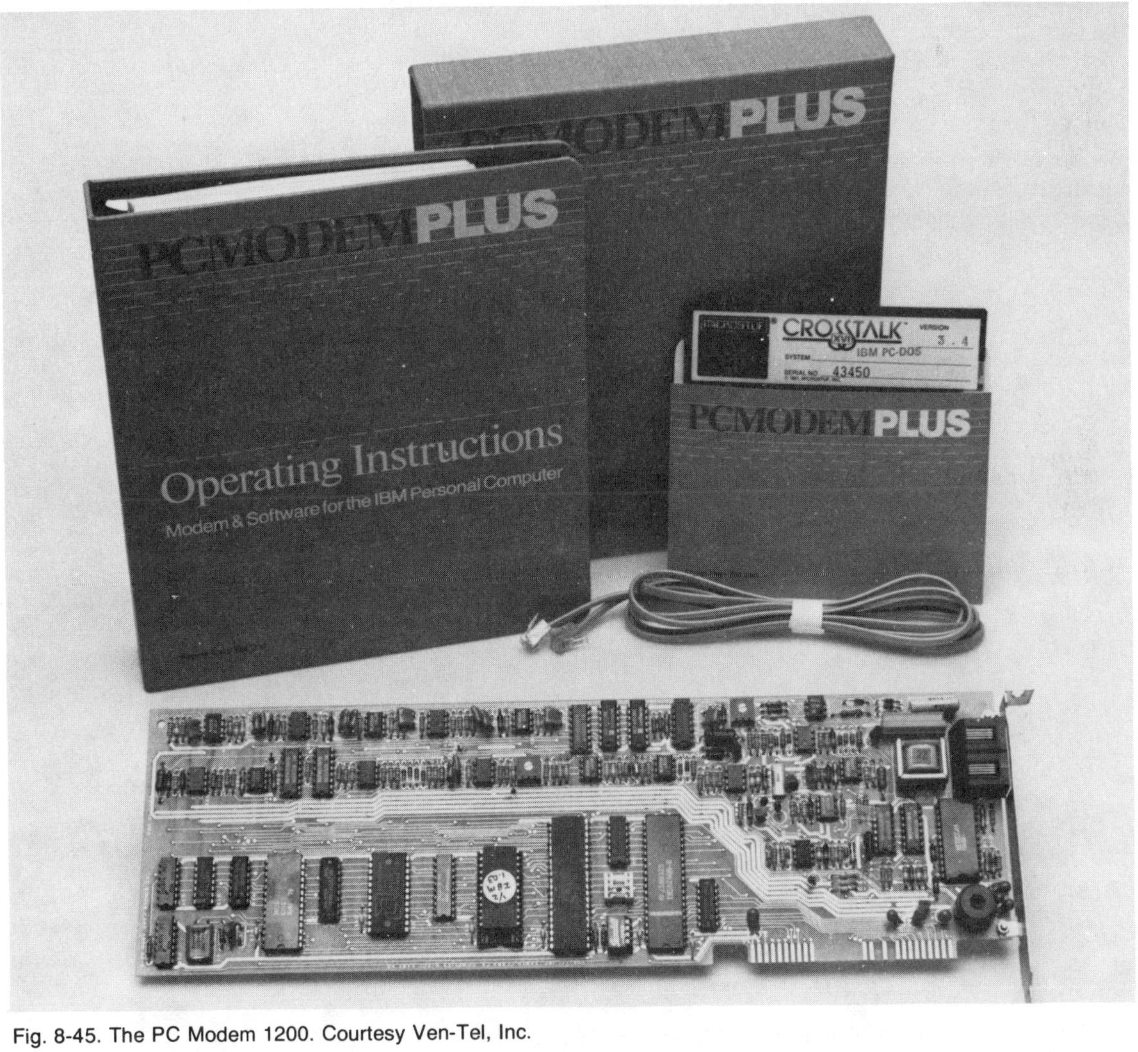

Fig. 8-45. The PC Modem 1200. Courtesy Ven-Tel, Inc.

software, including Smartcom II and PC Talk III.

Discussion. The documentation provides a thorough description for installing the modem into the computer. A set of DIP switches on the card provides the option of automatically answering incoming calls.

Many of the functions of the modem are actually the result of the software Crosstalk (see Chapter 9). The modem itself is fully programmable. It assumes one of two states: the local command mode, where commands controlling the activities are transmitted to the modem; or the on-line mode, where data is actually transferred through the modem. Most of the commands are transmitted by the software automatically, in response to files you create in advance. A special command mode allows you to transmit some of these control characters directly from the keyboard.

Files can be created for executing the auto-dial feature. Additional commands necessary for logging on to a system can be included in the file, as well as specific RS-232C parameters for that number. A logon sequence can recognize responses from the other system and response accordingly.

Commands for auto-dial provide for touch-tone dialing, pulse-dialing, pausing for two seconds, repeating dialing instructions, and returning to command mode. Commands for answering include answering after one ring, answering after five rings, and answering immediately.

Through Crosstalk, your terminal can emulate a number of other types of terminals for accessing specific computer systems. Emulated terminals include Televideo 910/920, IBM 3101 ASCII terminal, Adds Viewpoint, and DEC VT-100.

1200 Plus

BAUD: 1200/300
COMPUTER: Any RS-232C-compatible system
TYPE: Bell 212A/103

Fig. 8-46. The 1200 Plus modem. Courtesy Ven-Tel, Inc.

SOFTWARE: Not included
AUTO ANSWER: Yes
AUTO DIAL: Yes (pulse- or tone-dial)
PHONE: RJ11 (modular) connection

Features. The 1200 Plus is a 1200/300-baud, stand-alone, external modem (Fig. 8-46). A standard female DB25 connector on the back panel allows connection to any RS-232C-compatible system. The modem includes two RJ11 jacks that allow you to leave a telephone connected for normal use when the computer is not on-line.

The modem includes auto-dial features, using either pulse- or tone-dial systems. A built-in speaker allows you to monitor the progress of calls.

Discussion. Four external switches configure for various applications: display result codes in English; send the result codes to your computer screen; echo characters while in the command mode so that the commands entered will be displayed on the screen; and answer the phone automatically.

The four internal switches make other configurations: enable the speaker, enable the command recognition, force the DTR signal high (if your computer does not support the DTR line), and force the DSR line high (if your computer does not support the DSR line).

LEDs on the front panel indicate various states: high-speed (1200 baud) mode; auto-answer mode, carrier-detect (after contact with the distant computer), off-hook (to the phone line), receive data signal, send data signal, terminal ready (to send or receive), and modem ready (when on-line or during an analog loopback test).

The 1200 Plus operates in one of two states: the local command mode, where commands controlling the activity of the modem are sent; and the on-line mode, when data is transmitted through the modem. In the local command mode, one- and two-character commands can be sent from the terminal to the modem. These commands can be used to dial numbers, repeat commands, wait for dial tones, and return to the command mode. For example, entering "AT DP 767-8900" will enter the command mode, pulse-dial the number, and return for the next command.

An additional command set can be used to program responses in the auto-answer mode. The modem can be programmed to establish contact for data transfer by answering the phone, either immediately or after waiting for a specified number or rings.

A third set of commands configure the various RS-232C parameters. For example, the command "F1" will set the modem for full-duplex operation.

A number of commands implement the same features as those established through the eight switch settings. These commands have priority and will overpower the switch settings.

A final set of commands can be used to perform a loopback test, in either the originate or answer mode. After the command is issued, characters entered from the keyboard go to the 1200 Plus transmitter section and, instead of going out over the phone lines, reroute to the receiver section. The appearance of the characters on the terminal screen verifies the modem's working condition.

The modem, by using the same standard command sets, should work with most advanced communications software, including Crosstalk and Smartcom II.

HP 150 Internal Modem

BAUD: 1200/300
COMPUTER: HP 150
TYPE: Bell 212A/103
SOFTWARE: Not included
AUTO ANSWER: Yes
AUTO DIAL: Yes (pulse- or tone-dial)
PHONE: RJ11 (modular) connection

Features. The HP 150 Internal Modem inserts into any of the full-size expansion slots (Fig. 8-47). The modem includes two RJ11 jacks that allow you to leave a telephone connected for normal use when the computer is not on-line.

The modem includes auto-dial features, using either pulse- or tone-dial systems. A built-in speaker allows you to monitor the progress of calls. The modem also includes an auto-answer feature, allowing callers to establish contact and leave mes-

Fig. 8-47. The HP 150 internal modem. Courtesy Ven-Tel, Inc.

sages on an unattended system.

The modem is fully compatible with Hewlett-Packard software, such as Personal Card File and Series 100 DSN/Link. Commands can be entered from the terminal mode of the HP 150 without additional software.

Discussion. The documentation provides thorough instructions for installing the modem into either of the two slots within the computer. A set of DIP switches on the card provides the option of automatically answering incoming calls.

The modem itself is fully programmable. It assumes one of two states: the local command mode, where commands controlling the activities are transmitted to the modem; or the on-line mode, where data is actually transferred through the modem. Most of the commands are transmitted automatically by software, such as DSN/Link or Personal Card File. In the terminal mode, commands can instead be entered manually.

Commands for auto-dial provide for touch-tone dialing, pulse-tone dialing, pausing for two seconds, repeating dialing instruction, and returning to command mode. Commands for answering include: answering after one ring, answering after five rings, and answering immediately. The command results display on the screen. For example, after dialing a number, the display might be RING. Then, after detection of the carrier signal, the display might be CONNECT.

The HP 150 permits you to configure the various RS-232C parameters from the keyboard. Additional parameters, such as the length of time to wait for a carrier signal before disconnecting, are configured by entering values into various registers. The documentation discusses this process in detail.

In an internal printer is installed in the HP 150, a record of information received through the modem can be printed directly. The modem automatically senses the speed of transmission from the other modem and adjusts its rate accordingly. When working with a particular program, however, the HP 150 may need some adjustment for data-transmission speed.

VISIONARY ELECTRONICS

141 Parker Ave.
San Francisco, CA 94118
(415) 751-8811

Visionary Electronics is a relatively new company marketing two rather unique modems. The products are sold mainly through direct order with the company.

The Visionary 100 is warranted in its entirety against any defects of material or workmanship which develop for any reason whatsoever, except abuse, within a period of 120 days.

Models reviewed and suggested retail prices:

Visionary 100	$595.00
Visionary 1200	$795.00

Visionary 100

BAUD: 300
COMPUTER: Any RS-232C-compatible system
TYPE: Bell 103
SOFTWARE: Not included
AUTO ANSWER: Yes
AUTO DIAL: Yes (pulse- or tone-dial)
PHONE: RJ11 (modular) connection

Features. The Visionary 100 is a 300-baud "intelligent" modem, especially well suited for electronic mail. An internal command processor recognizes and stores lists of commands. The Visionary 100 responds to English commands, such as DIAL, SEND, AWAIT, CAPTURE, PRINT, ERASE, MONITOR AND FORGET, and MEMORY. Entering commands from the keyboard of the computer, the modem can be programmed to automatically dial a number at a preset time, logon with appropriate responses, download information from the remote service, and save the data for later use. Once programmed, the computer can be turned off, and even disconnected from the modem. The Visionary 100, using an internal clock, will automatically perform the programmed entries at the correct time.

The Visionary 100 can be programmed to answer the phone and receive and store incoming messages. Again, the computer need not be connected at the time.

Additional software is required to establish an RS-232C exchange of data between the Visionary 100 and the connected computer.

The Visionary 100 requires a 120-volt ac source, but it includes rechargeable NICad batteries backing up memory internal to the modem.

Discussion. The Visionary 100 behaves essentially like a dedicated communications computer, requiring only a keyboard and some sort of output device for viewing information.

In an interactive mode, the modem can be used to receive data directly from the keyboard of the computer and to transmit incoming data to the screen. In the program mode, information is obtained in a two-stage process. In the first stage, data is transferred, in either or both directions, between the modem and host computer. In the second stage, data is transferred between the modem and the computer.

In the program mode, phone numbers and messages, for distribution at various times, can be entered into the modem's memory. Redial (on busy or no answer) and auto logon features can be incorporated into English-language programs. Auto-answer features can be programmed to go into effect at preset times. Programmed functions will occur automatically at the designated time assigned within the program.

The Visionary 100 includes 2K memory that can be expanded up to 48K. The number of programs and amount of data that can be stored are limited only by the available memory.

Visionary 1200

BAUD: 1200/300
COMPUTER: Any RS-232C-compatible system
TYPE: Bell 212A/103
SOFTWARE: Not included
AUTO ANSWER: Yes
AUTO DIAL: Yes (pulse- or tone-dial)
PHONE: RJ11 (modular) connection

Features. The Visionary 1200 is an "intelligent" 1200/300-baud modem, especially well suited for electronic mail. An internal command processor gives the device its intelligence, providing features not found on most other modems. The Visionary 100 responds to English commands including, DIAL, SEND, AWAIT, CAPTURE, PRINT, ERASE, MONITOR AND FORGET, and MEMORY. Commands entered from the keyboard of the computer are stored into the modem's memory.

The modem can be programmed to automatically dial a number at a preset time, logon with appropriate responses, download information from the remote service, and save the data for later use. Once programmed, the computer can be turned off and even disconnected from the modem. The Visionary 1200 will automatically perform the programmed entries at the correct time.

The Visionary 1200 will automatically answer the phone and receive and store incoming messages. Again, the computer need not be connected at the time.

Additional software is required for establishing an RS-232C exchange of data between the Visionary 1200 and the connected computer. The Visionary 1200 requires a 120-volt ac source, but it includes rechargeable NICad batteries which back-up the memory.

Discussion. The Visionary 1200 behaves essentially like a dedicated communications computer, requiring only a keyboard and some sort of output device for viewing information. The modem can be used in an interactive mode, receiving data directly from the keyboard of the computer, and transmitting data to the screen.

In the program mode, information is obtained in a two-stage process. The first stage results in data being transferred, in either or both directions, between the modem and host computer. In the second stage, data is transferred between the modem and the computer.

In a program mode, dozens of phone numbers and messages, for distribution at various times, can be entered into the modem's memory. Redial (on busy or no answer) and auto logon features can be incorporated into English-language programs. Auto-answer features can be programmed to go into effect at preset times. All programmed functions will occur automatically.

A custom command can consolidate a number of commands. For example, the sequence of commands for accessing a remote system can be defined and equated to a special custom command. Calling up that command will execute the entire accessing sequence.

The AWAIT command can be used to logon to a system. It looks for an expected series of text. When that text arrives, the next command executes.

The Visionary 1200 comes with a 2K memory that can be expanded to 48K. The memory size determines the number of programs and the amount of data that the modem can store. The documentation is excellent, providing examples and a step-by-step analysis of each procedure.

9

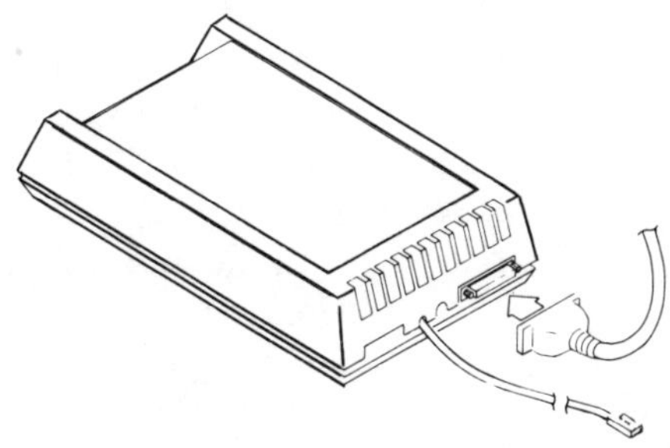

Communications Software Reviews

Reviews of communications software are listed alphabetically by company. Prices are as of 1984.

ALPHA SOFTWARE CONNECTION

30 B St.
Burlington, MA 01803
(617) 229-2924

Software reviewed and suggested retail price:
The Apple-IBM Connection $ 250.00

The Apple-IBM Connection

COMPUTER: Apple II, Apple II+, IBM PC
MODEM: Most modems, including Hayes-compatible modems

Features. The Apple-IBM Connection is a menu-driven communications package that includes two copies: one for the Apple II and one for the IBM PC. On the Apple side, the program requries a 64K Apple II+ or Apple II computer with one disk drive, and either a D.C. Hayes Micromodem II or a Mountain Hardware CPS Multifunction card. On the IBM PC side, the program requires a minimum of 96K of memory, one disk drive, and an asynchronous RS-232C interface.

The program can be used to transfer files between an Apple II and IBM PC computer, between two IBM computers, or between two Apple II computers. The program is designed for transferring only ASCII files; however, the communications package includes utilities for processing common files into ASCII files. VisiCalc, WordStar, Binary, and BASIC programs can be converted to ASCII before transmission and reconverted afterwords.

Programs can be transferred through modems or through a crossover cable, or mull modem.

The main menu displays a list of six modes: send, receive, send compare, receive compare, master, and slave. The same menu displays a list of three methods of linking: crossover cable, Hayes modem, and acoustic coupler. Also on the main menu is a list of baud rates—110 through 9600—and error checking methods: parity, checksum, and echo.

Discussion. In addition to the documentation, The Apple-IBM Connection includes a cassette tape which gives instructions for using the program. A slave option in the main menu configures the program for automatically answering a call and receiving messages on an unattended system.

Receive-compare and send-compare options in the main menu compare a transmitted or received file with a previously transferred file to check for errors in transmission. Parity and checksum error detection is standard and is always performed by the Apple-IBM Connection.

APPLE COMPUTER, INC.

20525 Mariani Ave.
Cupertino, CA 95014
(408) 996-1010

Software reviewed and suggested retail price:

Apple Access II	$150.00

Apple Access II

COMPUTER: Apple *II*e, Apple *II*c
MODEM: Apple modems

Features. Apple Access combines aspects of both the menu-driven and command-driven formats of the Apple *II*e and Apple *II*c computers. The program is menu-driven, and this determines the mode of operation; however, it has numerous commands with which to create specialized command files. The program is compatible with most modems, but is also requires an Apple Super Serial Card.

A main menu provides the following options: dial a service, terminal mode, set up communications, transmit a file, receive a file, utilities, help, and quit. A dialing directory of remote stations can be created. The directory can include a reference to a command file that provides information for logging onto a system.

Files can be transmitted or received with error correction, using XMODEM protocol, or without error correction, using XON/XOFF protocol. Files can be locked to prevent inadvertent deletion or alteration of a file.

Access II can be configured to emulate various terminals. It already has overlays for automatically emulating the ANSI, VT100, VT 52, and TTY terminals.

Discussion. You can create dialing directories, using almost any text editor. These directories can contain "dialing macros," which provide extended dialing commands. The directory can also reference command files that you create from over 55 commands. These command files can automatically logon to remote systems, transmit or receive files, set RS-232C parameters, emulate terminals and logoff—all between unattended systems.

Some systems transmit special characters which only a specific terminal can recognize. These characters confuse most other systems. For this reason, Access II includes a translation table that can be programmed to automatically change confusing characters. Translations already exist in the program for emulating TTY, VT52, VT100, and ANSI terminals.

BI-TECH ENTERPRISES, INC.

10 Carlough Rd.
Bohemia, NY 11716-2996
(516) 567-8155

Software reviewed and suggested retail prices:

Micro Link II Communications Package	$110.00
Uniterm	$ 49.95

Micro Link

COMPUTER: TRS-80 Model IV
MODEM: Most modems, including Hayes-compatible modems

Features. Micro Link is a command-driven communications package for the TRS-80 Model IV utilizing TRSDOS 6.x or DOS Plus IV. Various features of the modem are implemented by pressing the <clear> key in conjunction with a designated alphanumeric key. For example, <clear> <P> will turn the printer on or off.

A single logon message can be created and sent automatically to host systems which send out a Control E character after connection. A series of eight user buffers, of up to 63 characters each, can be created and transmitted with special keys. This feature can be used to initiate logon sequences, simplifying repetitive transmissions.

Through various commands, Micro Link can transmit or receive files with error correction, using XMODEM protocol. Data can be selectively captured in a 40K buffer and subsequently saved to diskette (without error correction). This process can be turned on and off to capture only selected data elements. Data can be routed to a printer, as well as to the screen, and turned on and off to print only desired text.

The program can be configured to set various RS-232C parameters necessary for exchanges with a particular system. These settings will activate automatically when the program is booted.

Binary files often include values which the modem misinterprets as control characters by a modem. For this reason, Micro Link includes a program that, before transmission, will convert binary files into ASCII files. After downloading, these files can be reconverted back to binary files.

Files can be transferred either in block format, where a buffer steadily empties, character by character, or in prompt format, where transmission pauses after every line until the receiving system is ready for the next line. Files can also be transferred with error detection, using the XMODEM protocol.

Incoming data can be routed to the printer for immediate printing. A single command will turn the printer on or off. Alternatively, the entire screen display can be printed one page at a time.

Uniterm

COMPUTER: TRS-80
MODEM: Lynx, Chatterbox, Microconnection

Features. Uniterm is a menu-driven communications package for the TRS-80 Model I and Model III which requires a minimum of one disk drive and 32K of memory.

The command mode makes available the following options: activate autobuffer, load or save binary files, close buffer, display or print buffer, exit to DOS, select half or full duplex, define initialization parameters, load ASCII file to buffer, change modem parameters, open and clear buffer, send buffer in prompt mode, send buffer in block format, transmit buffer, set screen width, type to buffer, and display command list.

A special "redefinition feature" allows you to redefine the output of keys. These keys can be programmed to transmit any ASCII value after first pressing the UP-arrow key. Four keys can be programmed to transmit logon information automatically.

Files can be captured into a buffer and then saved to diskette. Similarly, files can be moved to the buffer and then uploaded to a host system. Both ASCII and binary files can be transferred.

Discussion. The program can be configured to set various RS-232C parameters necessary for exchange with a particular system. These settings will activate automatically when the program is booted.

A 30K buffer is used as an intermediate storage area for uploading or downloading information. The buffer can be turned on and off to selectively capture information. When the buffer is full, a warning message displays. A buffer can be displayed, edited, and printed.

In data-transmission protocols, many computers use peculiar character which other computers may not recognize. Uniterm allows you to automatically change these characters before they confuse your computer.

COMMUNICATIONS RESEARCH GROUP, INC.

8939 Jefferson Highway
Baton Rouge, LA 70808
(504) 923-0888

Software reviewed and suggested retail price:

BLAST	$250.00

BLAST

COMPUTER: MS DOS systems
MODEM: Most asynchronous modems, including Bizcomp 1012, Vadic 212-PA, and Hayes Smartmodem

Features. BLAST (BLocked ASynchronous Transmission) is a menu-driven program compatible with msot intelligent modems. A translation table allows you to change character values to ensure compatibility with nearly all micro and mainframe computers.

The program displays a main menu with the following options: dial phone number, wait for incoming call and answer, do terminal function, do file-transfer function, do local-file management functions, hang up phone, display/edit configuration data, build/edit transfer command file, and return to operating system. Some selections will prompt for further information.

Incoming data can be selectively captured in a 24K buffer. The buffer data can be saved to disk automatically, using XON/XOFF protocol, or, upon request, the saving can be performed manually.

Files can be transferred using a special blocked-asynchronous-transfer protocol. This protocol provides error detection and correction and has a generally higher throughput than other types of transfer.

Command files with file-transfer commands can be created to automatically logon and access data from a remote system. Command files can also be used to automatically answer and service incoming calls.

Discussion. Through the display/edit configuration mode, dialing instructions for a particular system can be entered into a separate file. Calling the file will automatically enter dialing commands, the number, and additional RS-232C parameters necessary for connecting with a system. Dialing commands are limited to those recognized by your modem.

The build/edit file-transfer command mode allows the user to create files to automatically access information from a remote system or to configure a system to automatically answer and service incoming calls.

Selecting the file-transfer mode prompts the user for additional information, including the file name. A slave option allows the remote site to gain control of the local computer after a transfer has occurred.

A capture buffer can be turned on and off to receive selected parts of incoming data. The buffer can be reviewed or saved permanently on disk.

COMPUTER: APPLICATIONS, INC.

13300 SW 108 Street Circle
Miami, FL 33186
(305) 385-4277

Software reviewed and suggested retail price:

Apple-Link	$59.95

Apple-Link

COMPUTER: Apple II or Apple II+
MODEM: Hayes Micromodem II

Features. Apple-Link is a menu-driven program for data communications between two Apple systems. The program emphasizes the transfer of files between systems. The main menu displays the following options: auto dialer, converse, file transfer, and exit to BASIC.

The program includes uploading and downloading capabilities. Files transfer directly to and from diskette. Groups of files can transfer automatically between unattended systems.

During file transfers, each file sector is checked for transmission errors. Any error causes the sector to be retransmitted. An error summary displays at the end of a transmission.

Discussion. The auto-dialer option prompts the operator for a number. Once the number is entered from the keyboard, a carriage return starts the dialing of the number. If a connection does not occur within 30 seconds, the operator can try again or return to the main menu.

The converse mode allows the interactive exchange of data between two systems. Data entered through one system displays on the other's screen and visa versa.

In the transfer mode, the operator can either send or receive files. For sending, a list of files on diskette displays on the screen. Entering the character corresponding to a file, followed by a carriage return, sends that file. Any number of files can be sent. For receiving, the list of files on the remote computer's directory transfers to the local computer's screen. any of the files displayed can be selected and downloaded automatically. When transferring files, a clock display indicates the time to complete a transfer.

DIGITAL MARKETING

2363 Boulevard Circle
Walnut Creek, CA 94595
(415) 947-1000

Software reviewed and suggested retail price:

Micro Link II	$99.00

Micro Link II

COMPUTER: CP/M and MS DOS systems
MODEM: Most modems

Features. Micro Link II is a command-driven communications package. In the command mode, options and features are executed by entering one or two numerals from the keyboard. The commands and their corresponding numerals display in a series of four menus.

A sequence of commands, such as for dialing or logging on to a system, can be combined into a single command. Up to ten of these combination commands can be created to access various systems. Each sequence can include up to 60 characters.

Through various commands, Micro Link II can download data directly to disk. The downloading feature can be turned on and off to capture data selectively.

Micro Link II can send virtually any type of file. It can send an uploaded file, either complete or automatically edited to remove unwanted characters.

Discussion. In data-transmission protocols, many computers use special characters which other terminals may not recognize. Micro Link II allows you to filter out these characters before they confuse your computer.

Commands for transmitting data files allow the use of either XMODEM transfer, with error checking, or blind transfer, without error checking.

You can define special macro commands, called *phrases,* that will automatically transmit a string of characters. These phrases can automatically send, for example, a password, an identification number, or any string of text used often in your data communications. An initial setup phase prompts the user for various parameters involved in data transmissions. During this phase, you can create up to 10 phrases, each with an appropriate baud rate, parity, etc. These configurations can be saved and automatically implemented at any time.

Files can be transferred, using XMODEM protocol, for complete error detection and correction.

DIRECT AID, INC.

P.O. Box 4420
Boulder, CO 80306
(303) 442-8080

Software reviewed and suggested retail prices:

Direct Connect	$ 95.00
The Impersonator	$195.00

Direct Connect

COMPUTER: IBM PC
MODEM: Most modems, including Hayes-compatible modems

Features. Direct Connect is a command-driven communications package for the IBM PC/XT or compatibles. The program requires a minimum of 128K bytes of memory and at least a single diskette drive.

Direct Connect is divided into four modes: main I/O mode, for exchanging data with another computer; file-selection mode, for specifying the file to be exchanged; line-error and modem-status mode, for listing specific errors and current status; and edit-configuration mode, for specifying RS-232C parameters.

Various features and modes are implemented through a series of commands. Each command is executed by pressing the <alt> key in conjunction with the first letter of the command. For example, "Alt M" will put the program in the main menu mode.

The function keys can be programmed to automatically send a string of characters of commands. For example, these keys can be programmed to dial a number and to initiate a logon sequence. This function enables a user to customize a keyboard for specific and repetitive applications.

Through various commands, Direct Connect can transmit or receive files with error correction, using XMODEM protocol. Alternatively, the system can selectively capture data to a file (without error detection). The capturing can be turned on for only desired elements of incoming data. XON/XOFF protocol can be used when exchanging data with a remote system. The flow-control characters can be changed to accommodate any system. The DTR signal can be used to exchange data between two systems which are directly wired to each other.

Direct Connect includes parameters for emulating the Lear-Siegler ADM-3 terminal.

Discussion. Configuration files can be created to set various RS-232C parameters for exchanging data with a particular system. When accessing that system, these files can be called up to automatically set the appropriate values. Help screens can be called up, from almost anywhere in the program, to provide further information on a particular command or action.

A short start-up message can be created for display at the top of the screen whenever the program first starts up. The message can summarize configurations, and give telephone numbers, passwords, or any other information that might be useful to someone unfamiliar with the system.

Incoming data received through the modem can be routed to the printer for immediate printing. A single command will turn the printer on or off. Alternatively, the entire page displayed on the screen can be printed one page at a time.

The Impersonator

COMPUTER: IBM PC
MODEM: Most modems, including Hayes-compatible modems

Features. The Impersonator is a command-driven communications package for the IBM PC, XT or compatibles. The program requires a minimum of 192K bytes of memory and at least a single 320K diskette drive.

The Impersonator is divided into four modes: MAIN I/O mode, where data is exchanged with another computer; file-selection mode, where the file to be changed is specified; line-error and modem-status mode, where specific errors and current status are listed; and edit-configuration mode, where RS-232C parameters are specified.

The program recognizes 20 commands, each of which is executed by pressing the <alt> key in conjunction with the first letter of the command. For example, <alt> M will put the program in the MAIN menu mode.

The function keys can be programmed to automatically send a string of characters or commands. These keys can be programmed to dial a number and to initiate a logon sequence. This function enables a user to customize a keyboard for specific and repetitive applications.

Through various commands, The Impersonator can transmit or receive files with error correction, using XMODEM protocol. Alternatively, data can be selectively captured to a file (without error correction).

XON/XOFF protocol can be used when exchanging data with a remote system. The flow-control characters can be changed to accommodate any system. The DTR signal can be used to exchange data when two systems are wired directly to each other. The Impersonator includes parameters for emulating a number of terminals.

Discussion. In data-transmission protocols, many computers use peculiar characters which other computers may not recognize. The Impersonator allows you to filter these characters before they confuse your computer.

A set of preconfigured filters are included with the program to emulate ADM-3A, VT-52 and VT-100, Hazletine Esprit, Televideo 912, and IBM-3101 terminals.

Configuration files can be created to set various RS-232C parameters necessary for exchange with a particular system. When accessing that system, these files, which were saved to disk, can be called up to automatically set the appropriate values. Help screens can be called up from almost anywhere in the program to provide further information on a particular command or action.

The program includes a text editor for creating and editing messages and files. A short start-up message can be created that will display at the top of the screen whenever the program first starts up. This message can summarize configurations and give telephone numbers, passwords, and any other information that someone unfamiliar with the system might find useful.

Incoming data can be routed to the printer for immediate printing. A single command will turn the printer on or off. Alternatively, the entire screen display can be printed, one page at a time.

DYNAMIC MICROPROCESSOR ASSOCIATES, INC.

545 Fifth Ave.
New York, NY 10017
(212) 986-5872

Software reviewed and suggested retail price:
ASCOM $195.00

ASCOM

COMPUTER: CP/M, CP/M86, MS DOS, PC DOS systems

MODEM: Hayes-compatible modems

Features. ASCOM provides both menu-driven and command-driven formats. The menu-driven format leads the operator through a series of menus to a specific command. After selecting a command, the operator gets prompted for additional information. In the command mode, the operator can enter commands directly, along with additional necessary parameters, without having to work through a series of menus.

A BATCH mode allows operators to create a program from the various ASCOM commands. Executing that program causes ASCOM to process each command automatically. For example, a program of commands can dial, logon, and transfer data to another system.

An INDIRECT mode allows you to execute a program of ASCOM commands while under control of the operating system. When all of the commands have been processed, control returns to the operating system.

A REMOTE mode gives control of ASCOM to a caller. The caller can enter ASCOM commands and access information on an unattended system.

Through various commands, ASCOM can selectively save data received from another system. A special protocol, requiring a compatible system at the other end, can transfer files with complete error detection and correction.

Discussion. In data-transmission protocols, many computers use peculiar characters which other terminals may not recognize. ASCOM allows you to create translation tables which automatically convert these characters before they confuse your computer.

Data can be transferred to or from a remote system. Data received can be selectively saved into a buffer and then saved or printed.

Command files, which list any of the commands in ASCOM, can be created and saved to disk. These files can be used to dial a number, redial if necessary, logon to a remote system, access and download information, and disconnect. The command file can include RS-232C parameters and option settings.

Various commands enable the program to wait a predetermined amount of time for responses from a remote computer. A special time-out command allows the system to abort and disconnect if an interruption occurs for a specified amount of time. This command can be included in a command file for use when problems are encountered with an unattended system.

EXECS SOFTWARE

201 Waltham St.
Lexington, MA 02173
(617) 862-3170

Software reviewed and suggested retail price:
Termexec $79.95

TermExec

COMPUTER: Apple II+ or Apple *II*e
MODEM: Hayes Micromodem II, *II*e, Smartmodem, Zoom Networker, Novation Apple Cat, Anchor Signalman Mark XII, and Prometheus ProModem

Features. TermExec is a command-driven communications package. The various features and options are executed with one- or two-character entries from the keyboard. A help command displays a list of valid commands at the time. Keying in a question mark after the command will provide additional information about the form of the command. In a terminal mode, data from another computer displays as it arrives. Data that has scrolled off the screen can be scrolled back for review.

Keys can be assigned special meanings to automatically transmit character strings or TermExec commands. "Macro" commands can be created that will also automatically transmit text or commands.

Exec files can be created that contain long sequences of commands and text. A special LEARN command will capture and save all commands entered during an interactive session. Once saved, these commands can be called into action automatically.

Through various commands, TermExec can download data to disk via a buffer, or upload directly from disk. Transmission can be slowed down for slower remote systems. XMODEM protocol can be used in transferring data with a compatible system.

With an appropriate modem, TermExec can be configured to automatically answer incoming calls. This lets the caller use nearly all commands for exchanging data between the two systems.

TermExec is licensed to operate with Diversi-DOS, which permits file access three to nine times faster than with standard DOS. TermExec includes a free membership to DELPHI.

Discussion. Special files that define RS-232C parameters, options, screen-formatting instructions, and special key functions can be created and saved to disk. These files can then be called up at any time to automatically assume those settings. A similar type of file can be used to configure for a printer.

Keys on the keyboard can be assigned new meanings, either as a TermExec command or a string of text. Once pressed, these keys will automatically execute programmed commands or transmit strings of text. A single such key can transmit a password or an ID number, or issue a command to transfer a file. Referred to as *immediate-action macros,* these functions can be saved and loaded for use at any time. A short group of commands or text can be given a name and called into action by keying in the name.

For a lengthy sequence of commands, a separate file can be created that will automatically execute the commands when called into action. A phone number can be dialed, logon instructions transmitted, and data accessed, all in a single file. Delays can be included in the commands to wait for appropriate responses.

An EDITOR mode allows you to edit text files. Blocks of data can be moved, deleted, or copied. Various cursor controls allow full-screen editing.

HAWKEYE GRAPHIX

23914 Mobile St.
Canoga Park, CA 91307
(213) 348-7909

Software reviewed and suggested retail price:
Commx-Pac $150.00

Commx-Pac

COMPUTER: Configured for over 60 systems, including MS DOS, PC DOS, CP/M 80, CP/M 86, and Turbo DOS
MODEM: Most modems, including Hayes and PMMI

Features. Commx-Pac is a menu-driven communications package with configurations for over 60 different computers. A main menu provides the following options: local mode, terminal, copy from host, send to host, conversational, copy from Commx, send to Commx, create log file, copy from MODEM7, and exit.

Through various commands, Commx can transmit or receive files with error correction, using Commx or XMODEM protocols with compatible systems. Data can be selectively captured in a buffer and subsequently saved to diskette. This program supports the auto-answer capabilities of some modems, allowing callers with a proper password to access data from your system.

Discussion. Commx-Pac will automatically answer an incoming call and prompt for a password. Authorized callers can access data from the system. A user log keeps track of the numbers of successful and unsuccessful logons.

The program includes an electronic-mail subsystem, which can be programmed to call several unattended systems and exchange mail files. Macro files can be created to automatically transmit a sequence of characters or commands. Pressing a single key can initiate a logon sequence.

A dialing directory can be created with almost any text editor. Directory entries can contain commands which automatically configure the appropriate RS-232C parameters, dial a phone number, and provide logon data.

Some systems transmit special characters which only a specific terminal can recognize. Commx includes a translation table that can be programmed to automatically change confusing characters.

Incoming data can be routed to the printer for immediate printing. A single command will turn the printer on or off.

HAYES MICROCOMPUTER PRODUCTS, INC.

5923 Peachtree Industrial Blvd.
Norcross, GA 30092
(404) 441-1617

Software reviewed and suggested retail prices:

Smartcom I	Included with Micromodem IIe
Smartcom II	$149.00

Smartcom I

COMPUTER: Apple
MODEM: Micromodem IIe

Features. Smartcom I is a menu-driven program that is included with the purchase of Micromodem IIe for an Apple computer. The main menu presents a list of nine options: originate call, answer call, terminate call, create file, receive file, send file, list file, printer status, change parameters, and change configurations. Each selection produces a second menu of options, prompts for additional information, or does both.

To perform the various commands, the program enters two basic operating modes: local mode and on-line mode. In the local mode, configuration and parameter changes are made. In the on-line mode, information is exchanged between two systems. The on-line mode has two submodes: conversation mode and transfer mode.

Files can be transferred using the XON/XOFF protocol. Two systems with Hayes software can transfer files with error detection.

Discussion. Selection of CONFIGURATION and PARAMETERS from the main menu produces various options for configuring the computer, the printer, and the disk drive. The choices are saved on disk and automatically activated each time the program is loaded.

Selecting ORIGINATE CALL from the main menu produces a prompt for a phone number. The entry can have up to 35 characters, including dialing. A reference to one of three predefined numbers will automatically dial that number. Messages display the status of the call, such as DIALING or CONNECT. Also, the progress of the call can be heard through the modem's speaker.

Files can be transferred under three protocols: The XON/XOFF protocol, recognized by most commercial systems; the Hayes protocol, used between Hayes systems with error detection; and a line protocol that transfers one line of a file at a time.

Logon messages can be created and saved in a file. Transferring this file through the modem, one line at a time, can automatically dial a number and logon to a system.

Incoming data, displayed on the screen, can also be transferred directly to a printer. A character delay can be incorporated for transmissions to a slower device.

Smartcom II

COMPUTER: IBM PC and others
MODEM: Smartmodem 300, Smartmodem 1200

Features. Smartcom II is a menu-driven program with versions for the IBM PC and most other computers. It is designed to work with the Smartmodem 300 or Smartmodem 1200, also from Hayes. Six screens display the various options. The screens are: main menu, create, macro definition, parameter, configuration, and on-line. A status line provides a constant update of the program's activity. A help feature provides additional information about options.

The program automatically dials and logs onto a remote system. Telephone numbers can be stored in a communications directory or entered directly from the keyboard.

Smartcom II allows a sequence of commands or text to be stored together in a macro. You may define up to 26 macros for each system, include a set of RS-232C parameters for that system, and store the macros and parameters in a file that becomes part of the communications directory. The Smartcom II disk includes macros for popular information services.

Files can be transferred using the XON/XOFF protocol. Two systems with Hayes software can exchange data.

Discussion. Selection of configuration and parameters from the main menu presents various options for configuring the computer, the printer, and the disk drive. The various parameters are saved on disk and automatically initialized each time the program is loaded.

During logon, a typical data base will prompt for additional information. The responses can be inserted into macros. Macros are identified with a single character from A to Z. Pressing the macro key, followed by the macro-identification character, will automatically transmit the macro. Macro Z transmits automatically after a number is dialed and a connection is made. If the logon sequence is placed in the Z macro, Smartcom II will automatically logon to a system; you won't need to make any keyboard entry.

Files can be transferred under three protocols: The XON/XOFF protocol, recognized by most commercial systems; the Hayes protocol, used between Hayes systems with error detection; and a line protocol that transfers one line of a file at a time.

IDEASSOCIATES INC.

7 Oak Park Dr.
Bedford, MA 01730
(617) 275-4430

Software reviewed and suggested retail price:
IDEAComm 1200 Included with modem

IDEAComm 1200

COMPUTER: IBM PC
MODEM: IDEAComm modem

Features. You can create a phone directory of

up to 40 entries. Each entry can include a logon sequence and communication parameters. From a menu display of the numbers, the operator can simply key in a corresponding reference number to automatically dial the number. Messages inform the operator of the status of the call: DIALING, BUSY, etc.

Special modes enable uploading or downloading a file by name, with or without XMODEM protocol. The ON-LINE mode displays six options: Printer on/off, Disk Store on/off, Main Menu, XON/XOFF Enable, RCV XON/OFF, and on-line.

An auto-answer feature allows callers to establish contact and leave mesages on an unattended system. The IDEAComm 1200 will automatically display, in menu format, a list of options. The caller can look at the disk directory, leave a message, or upload or download a file. Callers can leave up to 15 messages on the system for the operator to view later.

An internal speaker allows you to monitor calls for dial tones, busy signals etc.

Discussion. A main menu lists the following options: set terminal characteristics, auto-originate mode, command mode, auto answer, return on-line, read messages, directory transmit file, receive file, and hang up. Some of these options lead to a second menu of options.

From the on-line mode, data can be selectively stored in a receiving buffer. Periodically, the buffer contents transfer to disk. Incoming data can also transfer directly to the printer. A special control can switch from a voice call in progress to a data-transmission call between computers.

A dialing directory of up to 40 names and numbers can be created for executing the auto-dial feature. Each entry of up to 30 digits can include commands for logging on to a system, as well as specific RS-232C parameters for that system. The logon sequence cannot recognize responses from the other system, but it can include delays for proper timing. Placing a call is as easy as entering the identification number and pressing return.

Of course, numbers not already in the directory can also be called directly from the keyboard, using the command mode. The operator is prompted for the phone number and type of dial tone (touch or pulse). After the operator keys in the number and presses return, the system dials the number automatically.

The modem can be configured for RS-232C parameters, using default settings for all transmissions. Individual parameters for each number in the phone directory can be preprogrammed.

LINDBERGH SYSTEMS, INC.

49 Beachmont St.
Worcester, MA 01609
(617) 852-0233

Software reviewed and suggested retail prices:

Omniterm	$95.00/$175.00
Omniterm II	$245.00

Omniterm

COMPUTER: TRS-80 Model 1 or III
MODEM: Radio Shack's RS-232C interface or a Microconnection "Port 208"-type modem

Features. Omniterm is a menu-driven program compatible with most intelligent modems. A translation table allows you to change character values to ensure compatibility with nearly all micro and mainframe computers.

The program displays a command menu with options for printer, screen formatting, carriage-return suppression, line-feed suppression, duplex, echo, carriage return after line feed, input to buffer, output from buffer, system commands, modify translation table, charge UART settings, send control-A and quit, send "AT" symbol and quit, scroll back display, zero real-time clock, fill buffer from disk, and save buffer to disk.

Selecting the system-commands option produces a second menu with additional options: auto signon and auto dial, file transfer, and saving or loading omniterm configurations.

Files can be transferred between a disk and buffer, or between buffer and a remote computer. A special XMODEM transfer process allows files of unlimited length to be transferred to or from a compatible system that has the XMODEM protocol. The status of each block of data transferred is displayed on the screen.

Discussion. Many computers include peculiar characters in their protocol for transferring data. These characters are recognized by their own terminals but confuse other systems. Omniterm allows you "catch" these characters and automatically change them to something recognized by your system. Using transformation tables, any character can be redefined so as to achieve complete compatibility with nearly all mainframe or micro computers.

The printer mode, accessed from the main menu, can toggle a printer on or off and also provide a 2048-byte intermediate buffer to accommodate slower printers.

An automatic dial and logon mode allows you to enter and save to a file dialing commands, the phone number, and additional RS-232C parameters. A logon sequence can include commands that halt the sequence for a specified amount of time. These built-in pauses can be used with a complicated procedure for logging on to a remote computer which prompts for information and waits for responses. The logon file can be called up and executed automatically.

To send a disk file to another system, the file must first be placed in a memory buffer. This feature limits the size of the file to the amount of memory. The size and amount of remaining memory displays in the bottom right line of the screen. The contents of the buffer can be examined at any time.

Files can also be transferred using the XMODEM protocol, if connected to a system with the same capability. Files transferred in this way are limited only to the amount of free disk space.

Text that has scrolled off the screen can be scrolled back again for review. Text from systems with an 80-column display can be reformatted automatically to accommodate the TRS-80 64-column display.

Omniterm2

COMPUTER: IBM PC and compatibles

MODEM: Most intelligent modems, including Hayes Smartmodem 1200, Racal-Vadic VA3450 series, Novation SMART-CAT, Concord Data Systems 224 AutoDial, and U.S. Robotics AutoDial 212

Features. Omniterm2 is menu-driven program compatible with most intelligent modems. A translation table allows you to change character values to ensure compatibility with nearly all micro and mainframe computers.

The program displays a main menu with options for terminal emulation, disk functions, display functions, printer functions, automatic dial and logon, macro keys and translation tables, RS-232C parameters, save and load settings, pause-accept definition change, pause-request definition change, sending files, and receiving files. Each menu selection produces a secondary menu of specific options.

The main menu also displays the current RS-232C parameters, as well as the status of the RTS, DTR, CTS, DSR, CD, and RI lines. During transmissions, the screen displays the number of parity, framing, and overrun errors.

Using the XMODEM protocol, files can be transferred directly to and from disk without the need of main memory buffers. File lengths are limited only by disk space. Data which has moved off the screen can be scrolled back for reviewing, and data can be searched for keywords or characters (1-66).

Discussion. Many computers include peculiar characters in their protocol for transferring data. These characters are recognized by their own terminals but confuse other systems. Omniterm2 allows you to automatically change these characters to something recognized by your system. Using transformation tables, you can redefine any character so as to achieve complete compatibility with nearly all mainframe or micro computers. A terminal-emulation mode emulates either a DEC VT100, a DEC VT52, or a TTY "dumb" terminal.

A disk-function mode, accessed from the main menu, allows you to view a file directory, erase files, view and search files, and change default drives. A display-function mode, accessed from the main menu, permits a color mode, for color monitors, and an expanded-menu mode, for viewing RS-232C parameters. A printer mode, accessed from the main menu, can toggle a printer on or off, change adapter numbers, or send special characters to the printer.

An automatic-dial and logon mode prompts you for modem commands, the phone number, and any logon information. The logon sequence can include commands which instruct the computer to wait for a prompt before proceeding. Most mainframe computers will prompt a terminal for information concerning account numbers and passwords. Omniterm2 will wait for the questions and then send the answers.

A file-receive mode, accessed from the main menu, allows you to set up a file to receive data. Data received from a remote computer displays on the screen and is saved in the file. If desired, a function key allows you to stop receiving data into the file. In the file-send mode, entire files can be transferred from disk, or individual characters can be sent as they are entered through the keyboard. Both sending and receiving file transfers can also use the XMODEM protocol, providing both systems have the capability.

Special keys can be defined that will automatically transmit a programmed string of characters. These macro keys can automatically send a password, an identification number, or any string of text used often in your data communications.

A pause-accept definition-change mode, accessed from the main menu, allows you to implement special ways to delay the transmission of data. For example, if the remote system is using XON/XOFF commands, your system can use those commands also, when necessary, to interrupt the transmission data. Other delay methods include: waiting for characters to be echoed back, waiting for a special prompt from the remote system, pausing between characters, or pausing between lines. Omniterm2 can use any of these methods in both transmitting and receiving modes.

When used with a modem having auto-answer capability, Omniterm2 can be configured to answer incoming calls, display a welcome message, prompt for a password, verify the password, and display a menu to the caller. The caller will then be able to send or receive files, obtain directory listings, print data through the host computer, and chat with the host operator.

MARK OF THE UNICORN

222 Third St.
Cambridge, MA 02142
(617) 576-2760

Software reviewed and suggested retail price:
PC/InterComm $99.00

PC /InterComm

COMPUTER: IBM PC and compatibles
MODEM: Most modems

Features. PC/InterComm is a menu-driven program compatible with most modems. It emulates the DEC VT100 terminal, allowing use with any software designed to control that terminal. The program displays a main menu for port options, interface options, style options, file options, file functions, other functions, tab settings, and explain keyboard. Each menu selection produces a secondary menu of specific options. A help function provides additional information about options.

Some of the most common menu options can be executed directly with assigned keys. Each function key can be programmed to send a string of up to 23 characters. A set of configuration values, including the function keys, can be saved to a file on diskette.

Files can be uploaded or downloaded, via a buffer, to disk storage. Data can be reviewed, even if it is not being saved to disk. When downloading, transferring data between buffer and disk can be controlled manually or automatically. When uploading, automatic delays can be included for slower receiving systems. Transmitted files can include binary or ASCII text. XMODEM protocol can be used with compatible systems.

Data can be printed as it arrives, or a screen dump can print only what is on the screen at the time.

Discussion. The PC/InterComm emulates the VT102, VT100, or VT52 terminals. Control signals that are used between computers and these terminals are processed appropriately.

A file-options mode, accessed from the main menu, allows you to set up a file for receiving data. Data from a remote computer displays on the screen as it arrives. Alternatively, the data can be saved to disk via a temporary buffer. With systems recognizing standard flow-control characters (XON/XOFF), the remote system is periodically told to delay transfer while data move from buffer to disk. With other systems, data can be manually transferred when the buffer is full. PC/InterComm is capable of transferring ASCII or binary files. The files can be transferred using XMODEM protocol or PC/InterComm protocol.

You can define up to 30 special keys that will automatically transmit a programmed string of characters. These keys can be used to automatically send a password, an identification number, or any string of text used often in your data communications. An answer-back message can be programmed and sent automatically when prompted by the host computer. The entire configuration can be saved to disk and recalled at any time.

The screen can be formatted to automatically wrap words that extend past 80 columns. The screen can display bold, inverse, and underscore. A status line can be used to display the last 79 characters, or the current time, and the length of time you have been running the program.

MICROSTUF, INC.

1845 The Exchange
Atlanta, GA 30339
(404) 998-3998

Software reviewed and suggested retail price:

Crosstalk	$195.00

Crosstalk

COMPUTER: IBM PC, XT and compatibles

MODEM: Most modems

Features. Crosstalk is a command-driven communications package. The various features and options are executed with one- or two-character entries from the keyboard. A status screen displays a list of commands, as well as current settings on major options. Some commands cause prompts for additional information. For example, the command "NU" produces the prompt: "Enter number to dial." In the STATUS mode, commands can be entered for the display of additional help screens. In the TERMINAL mode, data from another computer displays as it arrives. Several of the keys execute special commands at any time. For example, one key switches between TERMINAL and STATUS modes, and another key interrupts the TERMINAL mode to issue commands.

The function keys on the keyboard can be assigned new meanings, either as a Crosstalk command or as a string of text. Once pressed, these keys will automatically execute preprogrammed commands or transmit a string of text. For example, a single key can transmit a password or an identification number, or issue a command to transfer a file. The assigned functions can be saved in a command file and, at any time, loaded into memory.

Through various commands, Crosstalk can transfer data to your system's main memory or directly to disk. You can selectively capture data in memory, and, if desired, copy it to disk. When received to disk, the program writes the captured data every 1024 bytes.

Special command files can be created to execute a series of Crosstalk commands. With these files, dialing, logging on, and transferring data, for example, can be performed automatically.

With an appropriate modem, Crosstalk can be configured to automatically answer incoming calls. It greets callers with a preprogrammed message and can require them to supply a password. On verifying the password, Crosstalk grants various degrees of access. For example, it might allow full access to the system or prohibit the writing of files.

Crosstalk can emulate a number of terminals, including the IBM 3101, Televideo 910/920 series,

DEC VT-100, and Adds Viewpoint.

Discussion. In data-transmission protocols, some computers use peculiar characters which other terminals may not recognize. Crosstalk allows you to automatically filter out these characters before they confuse your computer.

Data can be transferred to or from a remote system, and received data can be selectively saved. A special retro-capture feature allows Crosstalk to save text that has already scrolled off the screen. A search feature allows you to search through the captured data for particular text. For transmitting to systems which operate at a slower rate, Crosstalk can insert automatic delays. Between systems using Crosstalk or Transporter, files can be transferred with error checking; erroneous data gets retransmitted and displayed.

Command files, listing any of the commands in Crosstalk, can be created and saved to disk. These files can be used to dial a number, redial if necessary, logon to a remote system, access and download information, and disconnect. A command file can include RS-232C parameters and option settings. Command files can also call up additional SCRIPT files. SCRIPT files can also contain commands: an "IF" command provides decision-making capabilities; a "JUMP" command begins the performing of commands in the SCRIPT file, and a "WHEN" command tells Crosstalk to perform an action upon receipt of a certain string of text.

Various commands enable the program to wait a predetermined amount of time for responses from a remote computer or, on recognizing characters from the modem, make a conditional response. Printing commands can be used to print files or to print text as it appears on a screen.

MICRO-SYSTEMS SOFTWARE

4301-18 Oak Circle
Boca Raton, FL 33431
(305) 983-3390

Software reviewed and suggested retail price:
MicroTerm $79.95

MicroTerm

COMPUTER: Most computers, including the IBM PC
MODEM: Hayes Smartmodem, Radio Shack Modem II

Features. MicroTerm is a menu-driven communications package for the IBM PC. MicroTerm operates in two main modes: command mode, which accesses commands sent before communications; and communications mode, which accesses commands implemented after connection is made. In the command mode, a menu of 20 options are displayed. Entering the alphanumeric character next to an option implements that option. The lower portion of the main menu displays the current RS-232C parameters and available memory within the buffer.

Directories of phone numbers can be created and saved to diskette. Selecting "dial telephone" from the main menu will display the directory. Entering a letter associated with an entry will dial the phone number automatically.

The function keys can be programmed to transmit a string of up to 64 characters. The keys can then be used to logon to a system or transmit a password. Function-key values can be saved and loaded at any time to access specific host systems.

Through various commands, MicroTerm can transmit or receive files with error correction, using XFER protocol. The remote system must also have XFER capability, however. Data can also be selectively captured in a buffer and subsequently saved to diskette.

Discussion. Function keys can be created to automatically transmit a series of characters or commands. For example, pressing a single key can transmit a password or an identification number. The settings for the function keys, along with a set of RS-232C parameters, can be saved and later recalled. The program can be configured to automatically load one of these files when the program is booted up.

Some systems transmit special characters which only a specific terminal can recognize. These characters confuse most other systems. For this reason, MicroTerm includes a translation table that can be programmed to automatically change confusing characters.

Incoming data can be routed to the printer for immediate printing. A single command will turn the printer on or off.

MICROTECHNIC SOLUTIONS INC.

P.O. Box 2940
New Haven, CT 06515
(203) 389-8383

Software reviewed and suggested retail price:
Smart 64 $39.95

Smart 64

COMPUTER: Commodore 64
MODEM: Vic Modem, Hes Modem

Features. Smart 64 is a menu-driven communications program for the Commodore 64. An options screen displays the various modes. Various options prompt for additional information after selection. Programmable function keys will automatically transmit a user-ID, password, and strings of text.

Data received in the on-line mode can be selectively saved in a 28K buffer. Smart 64 will, either automatically or on request, transfer the buffer to disk. An extracting mode allows some editing of the buffer contents.

Discussion. A memory buffer is used to capture incoming data. The capture mode can be toggled on or off during transmission to selectively save data. Once in the buffer, the data may be edited, saved to disk, or printed.

Six of the function keys can be programmed to transmit strings of text, including a password and user ID. A seventh function key will prompt for a file to be uploaded. File names preceded by an X will translate the characters according to a Transmit table, which converts PET ASCII values into normal ASCII values. The table can be modified to establish compatibility with systems recognizing different codes. The various parameters which configure the printer, interface, and screen can be set and saved for later use.

Smart 64 can convert a BASIC program into an ASCII listing of the program stored as a sequential file. The program will reconvert the sequential BASIC file back to a program file.

Some of the disk commands can be accessed directly, including SCRATCH, RENAME, VALIDATE, NEW DIRECTORY, and COPY. The C-64 wedge commands can be used from the program.

MIDWEST MICRO INC.

311 West 72nd St.
Kansas City, MO 64114
(816) 333-7200

Software reviewed and suggested retail price:
SuperTerm $119.95

SuperTerm

COMPUTER: Commodore 64
MODEM: Vic Modem, Hes Modem

Features. SuperTerm allows users to access all modes of operation from the terminal mode. Most modes are accessed through a series of function keys. Options include: receiving buffer on/off, streaming on/off, logon mode on/off, communications menu, terminal/edit, uploading/downloading menu, command mode, and display/printer mode.

Files can be created that contain the text necessary for logging on. This file can be loaded into a special logon buffer and each string of text selectively transmitted to the host computer.

Files can be uploaded or downloaded, via a buffer, to disk storage. Captured data can be selectively saved to disk. SuperTerm supports the Steve Punter protocol, as well as Image-file transfer. Program files, sequential files, or Wordpro files can be transferred. Data can be printed as it arrives or after transmission.

Discussion. A memory buffer is used to capture incoming data. The capture mode can be toggled on or off during transmission to selectively save data. Once in the buffer, it may be edited, saved to disk, or printed. Data can be copied from disk into the buffer and then transmitted to another computer. The buffer can also be printed. Data transmitted from the buffer can be "throttled," slowing it down for slower systems.

The buffer can also be used for logon commands. An interactive session recorded in the buffer can be selectively edited. The edited version can be saved and recalled at another time for automatically transmitting logon text.

To transmit text, a cursor can be moved to a line of text. Each time the carriage return is pressed, the corresponding string of text is transmitted.

Some systems, configured to work with other terminals, send unwanted characters that confuse the Commodore 64. A special wildcard option allows these characters to be deleted automatically.

The various parameters which configure the printer, the interface, and the screen can be set and saved for later use.

A separate program, included with SuperTerm, converts a BASIC program into an ASCII listing of the program stored as a sequential file. The program will reconvert the sequential BASIC file back to a program file. Program files can be converted to Image files (used on CompuServe) and vice versa.

A function-key editor allows you to redefine the use of up to 52 keys with a four-character code. The new definitions can be saved and used at any time. When you press the keys, the new four-character code will be transmitted automatically.

The disk-command channel can be accessed from the terminal mode, permitting a directory display, renaming a file, and deleting a file. SuperTerm has the necessary configuration setting to automatically emulate the VT100 and VT 52 terminal.

MORROW, INC.

600 McCormick St.
San Leandro, CA 94577
(415) 430-1970

Software reviewed and suggested retail price:
ReachOut $299.00
with Morrow MM300 modem

ReachOut

COMPUTER: Morrow Micro Decision computers
MODEM: Morrow MM300

Features. ReachOut is a menu-driven communications program for Morrow Micro-Decision computers with the MM300 modem. It has four main menus: dialing, control mode, configuration, and protocol. Each selection produces a second menu of options.

Dialing can start either directly from the keyboard, or by selecting from a list of choices in a telephone-directory file. The directory can be set up to redial the number, if necessary, and to wait a specified amount of time before hanging up. Special "script" files can be included in the directory to execute a logon sequence.

The software includes features for transferring files either individually or as a batch. Between two systems using ReachOut, an error-detection system automatically retransmits data until it is received correctly.

Discussion. The control-mode menu controls the major modes of operation and is considered the main menu. The configuration menu permits you to change the operating characteristics of the modem. The protocol menu allows you to change from one type of error-checking protocol to another.

A dialing directory, created with any word processor, can contain up to 26 entries. Special sections of the directory, called *scripts*, can have logon sequences. Scripts can have up to 80 characters and include delays for waiting for dial tones, etc. Each of the characters in a script executes a particular command. A "wait for" command responds only to a particular prompt from another system. This feature is very useful in a logon sequence where you must ask various questions.

A logging mode allows you to download and save information received from a distant modem. The logging mode can be activated intermittently to save selected material.

Files can be uploaded individually or in batches. A batch file, which lists numerous other files, can be made with a word processing program. When the batch file is executed, ReachOut will automatically upload each of the files, one at a time.

The various modes of operation, such as originate and answer, and the various RS-232C parameters, such as baud and parity, are implemented through software control.

A protocol menu lists the ReachOut-to-ReachOut protocol, the "Christensen" protocol, the CP/M User's Group protocol, and the Echo-plex protocol. Each selection has its own error-detection scheme.

ReachOut includes two public-domain programs that compress binary files by 10 percent and text files by 40 percent. Once transmitted, the files can be expanded to their original form. Compressed files transmit faster.

All features of the Morrow Decision can be used normally. The information displayed on the screen can be sent to a printer or stored on diskette.

MYCROFT LABS, INC.

P.O. Box 6045
Tallahassee, FL 32314
(904) 385-1141

Software reviewed and suggested retail price:
MITE, MITE/MS, or MITE/86 $195.00

MITE, MITE /MS, or MITE /86

COMPUTER: Computers with CP/M, CP/M-86, or MS DOS operating systems. Preinstalled version available for over 60 computers, including IBM PC/XT, Apple II, and compatibles.

MODEM: Most intelligent modems, including Hayes Smartmodem 1200, Racal-Vadic VA212 and VS212, Novation SMART-CAT, Anchor Signalman Mark VII, Bizcomp 1012, Rixon 212A, Incomm 212A, and U.S. Robotics AutoDial 212.

Features. MITE is a menu-driven program compatible with most intelligent modems. During installation, you can select from a long list of compatible modems and automatically configure MITE for that device. Another list automatically configures the program for your specific type of computer.

The program displays a main menu with options for go start communications, hang up phone, enter site ID, load parameters from disk file, and save parameters on disk. The main menu also presents options for submenus: parameter, text-file upload, binary-file transfer, command processor, special features, option, text-file download, macro definition, character filter, and exit to operating system.

Files transmitted from another computer can be captured into an intermediate buffer. If the transmitting system recognizes the normal flow-control characters, the buffer will automatically transfer to disk each 2K bytes of data. Alternatively, you can manually interrupt transmission to transfer data between buffer and disk.

MITE has the facilities to transmit binary files in either direction, using the following protocols: MITE, XMODEM, CLINK, HAYES, IBM PC, and TEXT.

Discussion. In data-transmission protocols, many computers use special characters which other terminals may not recognize. MITE allows you to automatically filter out up to 10 of these characters before they confuse your computer.

The command processor mode allows access to some disk functions. In it, files can be copied, renamed, erased, listed, etc., depending on the version of the program.

A special-features menu mode, accessed from the main menu, enables you to select special colors to highlight the foreground or background. A parameter mode creates files for establishing contact with various systems. These files might contain RS-232C parameters, as well as dialing instructions for a particular number.

Special keys can be defined that will automatically transmit a programmed string of characters.

These macro keys can be used to automatically send a password, an identification number, or any string of text used often in your data communications. A macro-key definition can contain up to nine strings. The strings execute automatically and can include information for logging on to the remote system. Special characters can be included to wait a second, or wait for an echo, before proceeding.

The experienced user, who might find flipping between menus inconvenient, can implement local commands in the terminal mode. Terminal commands include: APPEND, which writes the captured data to disk file, without closing the file; CAPTURE ON or OFF, which toggles text-capture on or off and READ, which uploads a file from disk.

When used with a modem with auto-answer capability, MITE can be configured to answer incoming calls. The caller can then use any of the local or system commands to send or receive files, obtain directory listings, or leave messages.

NEW ERA TECHNOLOGIES, INC.

2025 Eye St. N.W. Suite 924
Washington, DC 20006
(202) 887-5440

Software reviewed and suggested retail prices:

Conexus	$624.00 (includes MIST)
Mighty Mail	$349.00 (includes MIST BASIC)
MIST/MIST+	$495.00 (includes data base)

Conexus

COMPUTER: MS DOS and CP/M systems
MODEM: Most modems

Features. Conexus is an application of MIST (Microcomputer Information Support Tool), also from New Era Technologies. (See MIST.) It enables your microcomputer to provide services normally associated with mainframe computers. It is a completely integrated electronic-mail, computer-teleconferencing, and bulletin-board system, capable of supporting up to 1,000 members. Conexus requires that you be running MIST, version 1.3 or later, and have several megabytes of free disk space on a hard-disk drive.

Conexus maintains a separate account with the name, password, and level of access for each eligible caller. After callers have logged on, Conexus displays a sign-on bulletin. It then displays a menu of possible bulletin boards or conferences.

Discussion. Conexus is designed to run a bulletin-board, conference, or electronic-mail service. The operator creates a membership list that contains identification numbers, passwords, and levels of access. An access level determines how much control a caller may have over the system. One level limits callers to read messages only. A regular level adds the ability to send messages and post items. A moderator level allows the caller to also delete conference items, to move any marker in any conference, and to remove participants from private conferences. An establisher level lets the caller establish and close the message system, conference, or bulletin board. A manager level can go to MS DOS to back up Conexus files. A gatekeeper level can add or remove members to or from Conexus.

The system is completely automatic; it does not rely on interaction by a system operator. In fact, the interaction of a caller with the system does not display on the screen, thus protecting the privacy of a caller's entries. The system operator can, however, link with a caller for an interactive exchange; the operator notifies the caller that the operator can view the session.

Mighty Mail

COMPUTER: IBM PC or Compaq
MODEM: Hayes Smartmodem 1200 or 1200B

Features. Mighty Mail is another tailored application of MIST. It automatically performs routine electronic-mail managment functions on ITT Dialcom. The SOURCE, Western Union Easylink, and MCI Mail. It requires 256K of RAM, two 360K disk drives, and MS DOS 2.0.

Mighty Mail's features include: automatic logging on, sending, collecting, filing and reading electronic messages, transmitting files to and from the host, and error tracking.

Users can prespecify data and the time for functions to be performed. For example, it can be programmed to automatically logon and to send or retrieve messages, without manual intervention.

Mighty Mail can also be accessed and controlled from a remote system. It directs the caller through menu-driven operations.

Discussion. From an initial menu, a user can select access, send, list, online, remote, pickup, time, inbox, filing, or xmit.

The access option allows the user to create codes and specify parameters for subsequent access to the system. For example, dialing numbers, baud rates, identification numbers, and passwords necessary to logon to the electronic-mail service are all entered through the access option.

The send option prompts with "TO?". Entering an electronic address produces another prompt: NOW or LATER? Selecting LATER produces a prompt for TIME and DATE. The send option will dial, connect, transmit any messages, and retrieve any messages in your electronic address.

The list option displays the current list of electronic-mail addresses, for viewing or modification. The online option allows the user to connect manually to an electronic-mail service.

The remote option prepares your system for half-duplex access by callers. The pickup option will automatically dial the mail service and retrieve any mail in your electronic address. The time option displays the date and time and dedicates your system to electronic mail. When the specified times arrive, mail will automatically be exchanged.

The inbox option allows you to review any received electronic messages. Messages can be viewed, edited, printed, copied, renamed or deleted.

MIST/MIST+

COMPUTER: CP/M and MS DOS systems
MODEM: Most modems

Features. MIST (Microcomputer Information Support Tool) is actually a programming language designed for telecommunications. Programs can be created from a list of commands, allowing your computer to serve as a host computer for others to call up and use. Menu-driven formats can be developed to create bulletin boards, electronic mail services, and data-base information services. New Era Technology provides tailored applications of MIST, including Conexus, which is used for electronic mail, computerized conferencing, and bulletin boards (see Conexus).

Although primarily designed for providing electronic mail, conferencing, or bulletin boards, MIST can also access other systems. Programs can be created to logon to and access from other systems automatically.

Its commands can handle five kinds of tasks: file maintenance, text editing and formatting, telecommunications, remote use, and data-base management. The tasks are performed in one of three modes of operation: connect, edit, or resource.

MIST's features are fully integrated. For example, you can display the directory of files while interacting with another computer and return to where you were on-line. You can do remote database searches. You can co-edit text with a remote user. You can also use the output of any command as the input for any other command.

MIST includes features for creating *routines,* sequences of commands executed by a single-word command. For example, a routine might contain all the commands for calling up and connecting with another computer.

MIST provides multiple levels of password protection for both local and remote modes of operation. It supports XMODEM protocol, as well as the sending and receiving of simple ASCII files.

Discussion. MIST is written for CP/M systems. MIST+ is written for MS DOS systems. MIST/MIST+ BASIC contains the programming language. MIST/MIST+ Complete contains the programming language and a text-oriented database-management subsystem called Resources. Resources lets the user create single-file data bases of up to 4 million characters. The data bases can have seven types of fields, of up to 12,000 characters each, in addition to keyword fields. Records are directly accessible via both indexed and string-searching techniques. Boolean logic is supported for performing AND, OR, and NOT combinations. Remote systems can call up a MIST-controlled system and, after providing the proper password, access information within the data base created through Resources.

Through various commands, you can modify, delete, display, print, rename, link, reference, and copy files in and out of the editor.

Through a line-oriented text editor, you can enter, edit, delete, insert, and link text. You can format text using document-formatting commands for margins, justification, centering, indenting, page numbering, tabbing, underlining, boldfacing, and spacing.

Telecommunications commands allow the user to interact with other computers, upload files from disk, download files to disk, and exchange files between MIST systems with error correction.

Remote-use commands allow remote terminals to interact with your system automatically. Seven layers of passwords can be installed for varying degrees of protection.

SANYO BUSINESS SYSTEMS

51 Joseph St.
Moonarchie, NJ 07074
(201) 440-9300

Software reviewed and suggested retail price:
Sanycom $105.00

Sanycom

COMPUTER: Sanyo Computers
MODEM: Most Hayes-compatible modems.

Features. Sanycom is a command-driven communications package. The various features and options are executed with one- or two-character entries from the keyboard. A menu command displays a list of valid commands at the time. In a terminal mode, data from another computer displays as it arrives.

Special files can be created that contain a sequence of commands and text. These files are created with a regular text editor, such as WordStar. Once loaded into memory, each command can be executed individually by pressing a special control key.

A terminal mode allows an interactive session with a remote computer. If the terminal command is followed by a file name, a file opens for saving onto disk all data received by the remote system. The save feature can be toggled on or off.

Files can be transferred with or without error checking. They can be sent either individually or in a series, or *batch*, operation. A directory of names and phone numbers can be created so that, by entering the names, the numbers dial automatically.

Discussion. A special file, which lists various commands or strings of characters, can be created and edited through a standard text editor, such as WordStar. Each command or string can be transmitted by pressing CTRL-W, followed by a letter from A to Z that corresponds to a particular entry.

A phone directory can also be created. Each number can include a name and one command. A CALL command, followed by a name on the list, will result in the number being dialed and the command sent to the receiving system.

Data received from a remote system is first captured in a memory buffer. Periodically, flow-control characters temporarily halt the incoming data so that the buffer data can be saved to disk. If the receiving system does not recognize the standard characters (X/ON X/OFF), some data may be lost. Received data can also be sent directly to a printer.

A batch option allows you to list a series of files for sequential transfer. A backup option will automatically make a backup when the name of a transferred file already exists.

SOFTWARE PUBLISHING CORPORATION

1901 Landings Dr.
Mountain View, CA 94043
(415) 962-8910

Software reviewed and suggested retail price:

PFS:ACCESS	$95.00

PFS:ACCESS

COMPUTER: PC DOS systems
MODEM: Bizcomp PC:Intellimodem, Hayes Smartmodem, IBM PC Jr. built-in, Novation Smart-Cat, POPCOM Model X100, Transend PC ModemCard, U.S. Robotics Password, and Ven-Tel Modem Plus

Features. PFS:ACCESS is a menu-driven communications package for PC DOS systems. It requires a minimum configuration of 128K RAM and one disk drive.

A main menu provides a list of nine services, including CompuServe, Dow Jones, EasyLink, The SOURCE, and MCI Mail. Four other services can be entered onto the list by the user. A "choose modem" option on the main menu presents a secondary menu of 13 modems from which to choose. Choosing one configures the program for that specific modem.

Using function keys, PFS:ACCESS can transmit or receive files. A security feature allows you to garble any files transmitted. Using a security code of your own making will render the transmitted text unreadable at the receiving end until after it's processed with the same code.

Discussion. Incoming data can be routed to the printer for immediate printing. A function key will turn the printer on or off.

The entire process of dialing a phone number and logging on to a service can be recorded while it is occurring. This record, along with the necessary transmission parameters, can be saved and used later to automatically connect with a service. The name of this service can be added to the main menu, along with the five prerecorded services. Selecting a service from the main menu will automatically make connection with the service.

A coding process, based upon the Data Encryption Standard of the National Bureau of Standards, allows you to code transmitted data to prevent unauthorized access to the transmission. The operator selects any eight characters to encode the transmission. Only those knowing the eight characters can unscramble the transmission.

Two word processors, PFS:REPORT and PFS:WRITE, both from Software Publishing Corp., are compatible with PFS:ACCESS.

SOUTHEASTERN SOFTWARE

6414 Derbyshire Dr.
New Orleans, LA 70126
(504) 246-8438

Software reviewed and suggested retail prices:

Data Capture PC	$120.00
Data Capture IIe	$ 90.00

Data Capture PC

COMPUTER: IBM PC
MODEM: Most modems, including Hayes-compatible modems

Features. Data Capture PC is a menu-driven communications package for the IBM PC. It implements dialing features with Hayes-compatible commands.

A main menu provides the following options: delete text from capture buffer, insert text into capture buffer, list capture buffer, print capture buffer, send capture buffer, file menu, options menu, toggle menu, and quit menu. Selecting file menu, options menu, or toggle menu produces secondary menus.

The function keys can be programmed to transmit dialing instructions or logon commands for a particular host. These settings, along with the RS-232C parameters necessary for communicating with a system, can be saved in an option file.

Data Capture can transmit or receive files directly to diskette, using XON/XOFF protocol. Alternatively, data can be selectively captured in a buffer and subsequently saved to diskette. Data can also be transmitted and selectively saved by line number.

Discussion. Function keys can be created to automatically transmit a series of characters or commands. For example, you can press a single key to dial a phone number. Another key will transmit a password. A third key can transmit an identification number. The settings for the function keys, along with a set of RS-232C parameters, can be saved to an option file. This file can be called from diskette to configure your computer for accessing a particular host.

Some systems transmit special characters which only a specific terminal can recognize. Data Capture includes a translation table that can be programmed to automatically change confusing characters.

Incoming data can be routed to the printer for immediate printing. A single command will turn the printer on or off.

The capture buffer can be used to create and edit text. Text can be deleted, added, or edited, one line at a time. The buffer has 28K bytes of memory on 128K systems, and 9K bytes on 64K systems.

Data Capture IIe

COMPUTER: Apple *II*e
MODEM: Most modems, including Hayes-compatible modems

Features. Data Capture IIe is a menu-driven communications package for the Apple *II*e. It implements dialing features with Hayes-compatible commands.

A main menu provides the following options: await phone call, dial phone call, edit line in capture buffer, hang up phone, insert lines into capture buffer, list capture buffer, print capture buffer, remove lines from capture buffer, quit, send capture buffer, toggle menu, options menu, and files menu. Selecting file menu, options menu, or toggle menu produces secondary menus.

The numeric keys (0-9) can be programmed to transmit dialing instructions or logon commands for a particular host. These settings, along with the RS-232C parameters necessary for communicating with a system, can be saved in a special option file.

Data Capture can transmit or receive files directly to diskette, using XON/XOFF protocol. Alternatively, data can be selectively captured in a buffer and subsequently saved to diskette. Data can also be transmitted and selectively saved by line number.

Discussion. Numeric keys can be programmed to automatically transmit a series of characters or commands. For example, you can press a single key to dial a phone number, transmit a password, or transmit an identification number. The special settings for the numeric keys, along with a set of RS-232C parameters, can be saved to an option file. This file can be called from diskette to automatically configure your computer for accessing a particular host.

Some systems transmit special characters which only a specific terminal can recognize. Data Capture includes a translation table that can be programmed to automatically change confusing characters.

Incoming data can be routed to the printer for immediate printing. A single command will turn the printer on or off.

The capture buffer can be used to create and edit text. Text can be deleted, added, or edited, one line at a time. The buffer size is 24K.

SUPERSOFT INC.

P.O. Box 1628
Champaign, IL 61820
(217) 359-2112

Software reviewed and suggested retail price:

Term II	$200.00

Term II

COMPUTER: CP/M Systems
MODEM: Most modems, including Hayes-compatible modems

Features. Term II is a command-driven communications package. The various features and options are executed with one- or two-character entries from the keyboard. Commands are entered from one of nine modes of operation: exit, terminal, conversation, send, receive, telephone subsystem, buffer load subsystem, editor, and CP/M command execution.

Through various commands, Term II can download or upload data, via a buffer, to disk. Files can also be exchanged with one of three protocols: The first protocol, used only between two Term II or XTERM systems, includes error detection. The second protocol does not include error detection, but is compatible with nearly any host computer. The third protocol, with automatic error detection, is compatible with MODEM2/XMODEM file-transfer protocol.

Discussion. A conversation mode displays data transmitted and data received on two separate lines, making interactive sessions between two users convenient and simple.

Between two systems using Term II or XTERM (also from Supersoft) files can be transferred with complete error detection. Errors automatically cause a block of data to be retransmitted. MODEM2/XMODEM, recognized by many CP/M users and used by a number of CP/M bulletin boards, is included with Term II as an alternative protocol for file transfer with error detection.

Terminal-emulation overlays can be created to enable a system to simulate particular types of terminals. Translation tables will automatically change the special characters used by some computers to those recognized by your computer. Existing overlays have been successfully tested with CDC Cyber 175 and 176s, DEC System 10, DEC System 20, Honeywell 6180.

Directories, of up to 10 phone numbers each, can be created and saved on disk to automatically dial a number. The directory entries can include comments.

TAU SYSTEMS CORPORATION

7115 Leesburg Pike #211
Falls Church, VA 22043
(703) 237-7200

Software reviewed and suggested retail price:

I/O Express $149.00

I/O Express

COMPUTER: PC DOS and CP/M systems
MODEM: Most modems, including Hayes, Signalman Mark XII, Bizcomp, Cermetek, Multi-Tech MT212AH, Novation, Penril, Racal-Vadic, Rixon, U.S. Robotics, and Visionary 1200

Features. I/O Express is a menu-driven communications package for PC DOS and CPM systems, including the IBM PC, KAYPRO, Morrow Micro Decision, and Televideo.

A main menu provides the following options: directory of files, establish batch list of files, channel directory, link a channel, transmit, receive, status display, exit to operating system, and end session.

A channel directory of up to 148 remote stations can be created. By entering a channel number for one of the remote stations, I/O Express will automatically set the appropriate RS-232C parameters, dial the phone number, and perform the necessary logon commands.

Through various commands, I/O Express can transmit or receive files with error correction, using I/O Express or XMODEM protocols with compatible systems. Data can be selectively captured in a buffer and subsequently saved to diskette. This process can occur with flow controls, or one line at a time.

The program supports the auto-answer capabilities of some modems, allowing callers with a proper password to access data from your system.

Discussion. To reduce transfer times, I/O Express can automatically compress data. A group of files, entered as a batch, can be transferred automatically between unattended systems.

Users can designate a word-wrap column. I/O Express will then automatically adjust all incoming or outgoing text to the corresponding width. Text captured in the buffer can be reviewed at a programmed rate, paused, erased, saved, or printed.

I/O Express will automatically answer an incoming call and prompt for a password. Authorized callers can access data from the system.

Incoming data can be routed to the printer for immediate printing. A single command will turn the printer on or off. Alternatively, the entire screen display can be printed one page at a time.

TRANSEND CORPORATION

2190 Paragon Dr.
San Jose, CA 95131
(408) 946-7400

Software reviewed and suggested retail price:

Transend PC	$189.00

Transend PC

COMPUTER: IBM PC, XT and compatibles
MODEM: Transend Modems or Hayes-compatible modems

Features. Transend is a menu-driven communications program emphasizing electronic-mail and file-transfer functions. The main menu displays a series of basket icons to represent the 24 functions performed by Transend. Keyboard controls move a cursor to a desired basket and function keys are used to perform a desired action on the basket.

Eight of the baskets have predefined functions: in, out, your PC, services, address, sent, reserved, and waste. The remaining baskets can be defined and labeled by the user.

When a basket is selected, it expands graphically to reveal an additional menu or list. Again, selections are made through cursor movement and function keys.

Discussion. After selecting your PC from the list of baskets, entries can be made to configure Transend for your computer and modem. Through the services basket, parameters necessary to access various systems, such as baud rate and parity, can be entered and saved. The parameters chosen for a particular system activate each time you connect with that system.

The address basket provides a list of names, phone numbers, and identification numbers that you can add, delete, and modify. Through the list, you can place a call automatically.

Transend is designed with electronic-mail services in mind. Messages can be sent and received by an unattended system. The program includes auto logon, multiple file transfer, multiple addressing, and address and message storage. Files can be transferred with XMODEM protocol for error detection.

A text editor facilitates the creation of messages. The text editor allows you to insert paragraphs, delete words, delete lines, restore lines, and move the cursor. Once created, messages are addressed and moved to the out basket. A list of messages can be transmitted, either immediately or at a later time. The system automatically dials the phone number, logs on, and transmits the messages. Transend will also automatically retrieve mail.

UNITED SOFTWARE INDUSTRIES

1880 Century Park, Suite 311
Los Angeles, CA 90067
(213) 556-2211

Software reviewed and suggested retail prices:

ASCII Express	$129.95
P-Term	$129.95
Z-Term	$149.95

ASCII Express

COMPUTER: Apple II, Apple II+, Apple *II*e, and Apple-compatible systems with DOS 3.3.
MODEM: Novation Apple Cat, Hayes Micromodem, Multi-Tech Modem II, Transend Apple Card, Mountain Computer CPS, ESI Lynx, and MPC Microconnection.

Features. ASCII Express is a command-driven communications package. In the command mode, the various features and options are executed with one- or two-character keyboard entries. Two menus list the various commands. A help command summons a brief description of each command.

Special keys can be programmed to automatically send a string of characters or commands. Called macros, these keys can be programmed to dial a number, initiate a logon sequence, upload a file, download a file, and hang up the phone. This feature enables a user to customize a keyboard for specific and repetitive applications.

Through various commands, ASCII Express can receive data into a disk file via an intermediate buffer of 18K bytes. If the remote system recognizes normal flow commands, data automatically transfers from buffer to disk; otherwise, the transfer can be accomplished manually. The buffer can also be used as a work area to selectively capture data or edit data from a disk file. ASCII Express can be configured to allow callers to connect with your system and access information.

Discussion. In data-transmission protocols, some computers use peculiar characters which other terminals may not recognize. ASCII Express allows you to automatically filter out these characters before they confuse your computer. Commands for transmitting data files allow the use of either XMODEM transfer, with error checking, or blind transfer, without error checking.

Special keys can be defined that will automatically transmit a string of characters that you have programmed in advance. These macro keys can be used to automatically send a password, an identification number, or any string of text used often in your data communications. Macro commands can also be designed to set various RS-232C parameters for exchanging data with a particular system. Macros can even reference and use other macros. A macro can be saved as a file on disk. A group of macro files serves as a sort of dialing directory. Selecting a particular macro can automatically dial a number, set the appropriate RS-232C parameters, or redial a number repeatedly until a connection is made.

File-control commands can be used to manipulate files on disk, such as copy, move, rename, list, or print.

When used with a modem that has auto-answer capability, ASCII Express can be configured to answer incoming calls. To control access to a system, it can require a caller to provide a password. It verifies the password automatically. The caller can then use any of the local or system commands to send or receive files, obtain directory listings, or leave messages. When the original caller hangs up, the local system will automatically disconnect and wait for the next call.

ASCII Express fully supports interrupts. While on line and receiving data, you can leave the terminal mode, issue a command, and return to terminal mode, without losing any data. Data is continually captured in an interrupt buffer and held for display.

P-Term

COMPUTER: Apple II, Apple II+, Apple *II*e, and Apple-compatible systems with Apple Pascal 1.1

MODEM: Novation Apple Cat, Hayes Micromodem, Multi-Tech Modem II, Transend Apple Card, Mountain Computer CPS, ESI Lynx, and MPC Microconnection

Features. P-Term is a command-driven communications package. In the command mode, the various features and options are executed with one- or two-character keyboard entries. Two menus list the various commands. A help command summons a brief description of each command.

Special keys can be programmed to automatically send a string of characters or commands. These macros can be programmed to dial a number, initiate a logon sequence, upload or download a file, and hang up the phone. This feature enables a user to customize a keyboard for specific and repetitive applications.

Through various commands, P-Term can receive data into a disk file via an intermediate buffer of 18K bytes. If the remote system recognizes normal flow commands, data automatically transfers from buffer to disk; otherwise, the transfer can be accomplished manually. The buffer can also be used as a work area to selectively capture data or edit data from a disk file. P-Term can be configured to allow callers to connect with your system and access information.

Discussion. In data-transmission protocols, some computers use peculiar characters which other terminals may not recognize. P-Term allows you to automatically filter out these characters before they confuse your computer.

Commands for transmitting data files allow the use of either XMODEM transfer, with error checking, or blind transfer, without error checking.

Special keys can be defined that will automatically transmit a string of characters that you have programmed in advance. These macro keys can be used to automatically send a password, an identification number, or any string of text used often in your data communications. Macro commands can also be designed to set various RS-232C parameters for exchanging data with a particular system. Macros can even reference and use other macros. A macro can be saved as a file on disk. A group of macro files serves as a sort of dialing directory. Selecting a particular macro can automatically dial a number, set the appropriate RS-232C parameters, or redial a number repeatedly until a connection is made.

File-control commands can be used to manipulate files on disk, such as copy, move, rename, list, or print.

When used with a modem that has auto-answer capability, P-term can be configured to answer incoming calls. To control access to a system, it can require a caller to provide a password. It verifies the password automatically. The caller can then use any of the local or system commands to send or receive files, obtain directory listings, or leave messages. When the caller hangs up, the local system will automatically disconnect and wait for the next call.

With the P-Term receiving data, you can leave the terminal mode, issue a command, and return to terminal mode, without losing any data. Data is continually captured in an interrupt buffer and held for display.

Z-Term

COMPUTER: Apple II, Apple II+, Apple *II*e, with Z80 Softcard
MODEM: Novation Apple Cat, Hayes Micromodem, Multi-Tech Modem II, Transend Apple Card, Mountain Computer CPS, ESI Lynx, and MPC Microconnection

Features. Z-Term is a command-driven communications package. In the command mode, the various features and options are executed with one- and two-character keyboard entries. Two menus list the various commands. A help command summons a brief description of each command.

Special keys can be programmed to automatically send a string of characters or commands. These macro keys can be programmed to dial a number, initiate a logon sequence, upload or download a file, and hang up the phone. This feature enables a user to customize a keyboard for specific and repetitive applications.

Through various commands, Z-Term can receive data into a disk file via an intermediate buffer of 18K bytes. If the remote system recognizes normal flow commands, data automatically transfers from buffer to disk; otherwise, the transfer can be accomplished manually. The buffer can also be used as a work area to selectively capture data or edit data from a disk file. Z-Term can be configured to allow callers to connect with your system and access information.

Discussion. In data-transmission protocols, some computers use peculiar characters which other terminals may not recognize. Z-Term allows you to automatically filter out these characters before they confuse your computer. Commands for transmitting data files allow the use of either XMODEM transfer, with error checking, or blind transfer, without error checking.

Special keys can be defined that will automatically transmit a string of characters that you have programmed in advance. These macro keys can be used to automatically send a password, identification number, or any string of text used often in your data communications. Macro commands can also be designed to set various RS-232C parameters for exchanging data with a particular system. Macros can even reference and use other macros. A macro can be saved as a file on disk. A group of macro files serves as a dialing directory. Selecting a particular macro can automatically dial a number, set the appropriate RS-232C parameters, or redial a number repeatedly until a connection is made.

File-control commands can be used to manipulate files on disk, such as copy, move, rename, list, or print.

When used with a modem that has auto-answer capability, Z-Term can be configured to answer incoming calls. To control access to a system, it can require a caller to provide a password. It verifies the password automatically. The caller can then use any of the local or system commands to send or receive files, obtain directory listings, or leave messages. When the caller hangs up, the local system will automatically disconnect and wait for the next call.

Z-Term fully supports interrupts. While on line and receiving data, you can leave the terminal mode, issue a command, and return to terminal mode, without losing any data. Data is continually captured in an interrupt buffer and held for display.

U.S. DIGITAL CORPORATION

5699-D S.E. International Way
Milwaukee, OR 97222
(503) 654-0668

Software reviewed and suggested retail price:
Copylink $99.95

Copylink

COMPUTER: A variety of computers, including IBM PC, Compaq, KAYPRO, Osborne I and II, Apple II (CP/M card), Molecular, Televideo, Morrow, Otrona, TRS 2000, North Star Horizon, and DEC Rainbow

MODEM: Most Bell 103- and 212-compatible or Hayes-type modems

Features. Copylink is a command-driven communications package. In the command mode, the various features and options are executed from the keyboard with one- or two-character entries. A help command displays the categories of commands. Other commands display the commands within each command category. The command mode displays a status line which gives the default drive, the free memory, and the baud rate.

A set command can give certain keys new meanings. Once implemented, these keys will automatically execute preprogrammed multiple commands. For example, a single key could initiate a logon sequence, upload a file, download a file, and hang up the phone. Set commands enable a user to customize a keyboard for specific and repetitive applications.

Command categories include: edit, additional, status, system, save, return, transfer, control, files, modem, terminal, and configure. Through various commands, Copylink can receive data into a disk file, via an intermediate buffer of 16K bytes. If the remote system recognizes normal flow commands, data automatically transfers from buffer to disk. Otherwise, the transfer can be accomplished manually. Copylink can be configured to enable callers to use various commands to access information.

Discussion. In data-transmission protocols, some computers use peculiar characters which other terminals may not recognize. Copylink allows you to automatically filter out these characters before they confuse your computer.

Commands for transmitting data files allow the use of XMODEM transfer, with error checking, or blind transfer, without error checking. Files can be sent from any disk. The use of wild-card indicators (? or *) allows the automatic transmission of groups of files.

Up to 10 special keys can be defined that will automatically transmit a string of characters. For example, macro-command keys can be used to automatically send a password, an identification number, or some text used often in your data communications. A macro string can include commands to control the transmission of data. For example, the speed of transfer can be altered to accommodate slower systems, or the system can be programmed to wait for the echo of transmitted characters.

A status line in the command mode displays a list of characters which relate to the operational modes currently in effect. File-control commands can be used to manipulate files on disk, such as copy, move, rename, list, or print.

Once Copylink has been configured to suit a particular set of hardware and software, a custom command will create a customized version of Copylink with those parameters. Autolog commands enable the program to wait a predetermined amount of time for responses from a remote computer or, on recognizing characters from the modem, make a conditional response.

You can create directories of phone numbers and names. Various commands can automatically dial a number from the directory, redialing the number repeatedly until making a connection; you simply provide the name and the directory.

When used with a modem with auto-answer capability, Copylink can be configured to answer incoming calls. A call can then use any of the local or system commands to send or receive files, obtain directory listings, or leave messages. When the original caller hangs up, the local system will automatically disconnect and wait for the next call.

U.S. ROBOTICS, INC.

1123 W. Washington Blvd.
Chicago, IL 60607
(312) 733-0497

Software reviewed and suggested retail price:

Telpack	$99.00

Telpac

COMPUTER: IBM PC, IBM PCjr.
MODEM: U.S. Robotics modems

Features. Telpac is a command-driven communications package with versions for the IBM PC and IBM PCjr. Telpac operates in two main modes: command mode, which accesses commands sent before communications; and communications mode, which accesses commands implemented after connection is made. The command mode gives access to five submodes: exit to subsystem, utility functions, telephone directory functions, help information, and communications mode.

Directories of phone numbers can be created and saved to diskette. Selecting an entry from the directory can automatically set RS-232C parameters, dial phone numbers, and enter logon messages.

Command files can be created to initiate a phone call, logon to a computer, access and download a file, and disconnect. The entire process can occur automatically from an unattended system.

Through various commands, Telpac can transmit or receive files with error correction, using XMODEM protocol or a user-defined protocol. Data can also be selectively captured in a buffer and subsequently saved to diskette. This program supports the auto-answer capabilities of some modems and allows callers with a proper password to access data from your system.

Discussion. Through various commands, Telpac will automatically answer an incoming call and prompt for a password. Authorized callers can access data from the system. A user log keeps track of the numbers of successful and unsuccessful logons.

Files made up of commands can be used to automatically access information from remote systems. These files can be used to dial a phone number and enter the correct responses to logon to a system. Additional commands can be used to exchange data, as in electronic mail. The same command file can be used to logoff. The operator can use XMODEM protocol, XON/XOFF protocol, or a user-defined data-transfer protocol.

VECTOR GRAPHICS, INC.

500 North Ventu Park Rd.
Thousand Oaks, CA 91320
(805) 499-5831

Software reviewed and suggested retail price:
Conect $150.00

Conect

COMPUTER: All Vector single- and multi-user systems
MODEM: Hayes Micromodem 100 and Smartmodem

Features. Conect is a command-driven communications package. Eight command modes are implemented by entering appropriate English commands. They include, send, receive, interactive, link, terminal, configure, execute, and help. A command menu displays the various commands on the screen and allows you to select the one desired. A help screen provides additional information about commands.

Using the send or receive command, Conect can transmit or receive CP/M files between two Vector systems running Conect software. Using the link mode, your CP/M disk can be accessed by a remote computer. Conect will answer an incoming call and treat the caller's commands as if they were entered from the console. It directs all output to the caller.

In the configure mode, RS-232C parameters can be assigned. A command line allows you to enter predefined commands. These parameters and commands can be saved on disk and automatically executed at any time.

Names and numbers can be stored in a directory. Entering the name causes Conect to automatically dial the corresponding number.

Discussion. Directories can be created with phone numbers and names. Entering the characters "#@," followed by a name in the directory, results in that number being dialed automatically.

The link mode works only between computers having 56K or more of memory. It answers calls and gives the remote computer complete access to the local disk.

The send command will automatically send an individual file to a receiving computer. A master file, containing a list of files, can be created to send multiple files automatically.

The interactive mode allows the interactive exchange of data between two devices. During transmission, all the information that appears on the screen can be captured in a buffer and saved to disk or printed.

The execute command implements parameters and commands created during configuration. A terminal mode configures the Vector system to emulate the Hazeltine 1500 terminal, allowing communication with computer systems that support that terminal.

WINDMILL SOFTWARE, INC.

2209 Leominister Dr.
Burlington, ON
Canada L7P 3W8
(416) 336-3353

Software reviewed and suggested retail price:
Video Link 88 $59.95

Video Link 88

COMPUTER: IBM PC and compatibles
MODEM: Most modems, including Hayes Micromodem

Features. Video Link 88 is a menu-driven program compatible with most modems. The program displays a main menu with options for terminal-mode operation, Hayes Smartmodem support, disk-related operations, change terminal specifications, and exit program. Function keys for use at any time while in the terminal mode will turn the printer on/off, save incoming data on disk on/off, save text screen on disk, start sending data from a disk file, stop sending data from a disk file, send a break signal, return to the terminal mode, return to the Smartmodem menu, return to the disk menu, and return to the main menu.

A directory of up to nine phone numbers can be created. Each number can include up to 25 characters, including any commands recognized by your modem. With Hayes-type modems, the directory-entered commands can redial, pause, etc. A single menu selection will automatically dial a number, using the commands included with the number.

A directory of nine transmission-parameter configurations can be created. Each configuration establishes baud rate, word length, parity, duplex, and protocol. A single menu selection, corresponding to a particular configuration, will automatically establish those parameters.

Discussion. A disk-related operations mode, accessed from the main menu, allows you to set up a file for receiving data. Data received from a remote computer displays on the screen as it arrives. A screen-save operation will save data displayed on the screen to a disk file. Additionally, incoming data can be selectively saved to disk with the special function keys. Disk files can be printed or displayed on the screen through additional options provided in the disk-operations mode.

Files can also be uploaded from disk. A 2-second delay can be automatically inserted at the end of each line to allow a distant computer time to receive data. Additionally, a keyboard control allows you to upload one line of data at a time.

An auto-answer mode allows an unattended system to automatically answer a call. Callers can leave messages on the screen, on the printer, or directly to disk.

ZOOM TELEPHONICS, INC.

207 South St.
Boston, MA 02111
(800) 631-3116

Software reviewed and suggested retail price:

Netmaster	Included with Networker modem

Netmaster

COMPUTER: Apple II/II+/*II*e and Franklin
MODEM: Networker

Features. Netmaster is a menu-driven program included with the Networker modem for use with systems a minimum of 48K of RAM and at least one disk drive. The program displays a main menu with the following options: converse, transmit/receive, printer access, disk access, configure, toggle upper/lowercase, toggle shiftlock, exit Netmaster, and help. Each option presents a secondary menu.

Data can be selectively received into or transmitted from a 40K buffer. The buffer can be reviewed, with a rapid or slow scroll, and individual screens can be printed on a compatible printer.

Netmaster has three means of transmitting files to other computers: Netmaster format, ASCII-converted hex format, and out of the buffer. Netmaster has two ways of receiving files: Netmaster format, or the buffer.

Discussion. In data-transmission protocols, many computers use special characters which other computers may not recognize. Netmaster lets you create translation tables to automatically change these characters for both sending and receiving data.

The converse mode is equivalent to what other systems usually call the terminal mode. Data entered through the keyboard transmits directly to the other system, and data received from the other system displays on the screen. Incoming data can also be saved selectively in the 40K buffer, for possible saving to disk, printing, or both.

A directory of up to 16 names and phone numbers can be created and displayed. Entering an item's label will automatically dial the appropriate number.

Files can be downloaded or uploaded, using a Netmaster protocol. The protocol can be used to transmit any file, regardless of its file type, with complete error detection and correction. For this type of file transfer, however, both transmitting and receiving systems must use Netmaster software.

Files can also be transmitted in an ASCII format. Each hex byte is translated into two ASCII characters, followed by a space.

Appendices

Appendix A

Product List

MODEM REVIEWS

ANCHOR AUTOMATION INC.
Volksmodem
Signalman Mark I
Signalman Mark III
Signalman Mark V
Signalman Mark XII

APPLE COMPUTER, INC.
Apple 300
Apple 1200

BUSINESS COMPUTER CORP. (BIZCOMP)
1022 Modem
1012 Modem
PC:IntelliModem

CERMETEK MICROELECTRONICS
Info-Mate 212A
Info-Mate 212PC

COMMODORE BUSINESS MACHINES INC.
VICModem (1600)
Automodem (1650)

COMMUNICATIONS AND PERIPHERAL MARKETING CORP.
MM-102
Automodem 1200A

COMPUTER COMMUNICATIONS SPECIALISTS, INC.
Audiomodem

CONCORD DATA SYSTEMS
224 Data Modem
224 Autodial Modem

DATASOLVERS, INC.
DS1200SMT

GANDALF TECHNOLOGIES INC.
Sam 212A

GENERAL DATACOMM, INC.
201C
212A/L
212A/ED

HAYES MICROCOMPUTER PRODUCTS, INC.
Smartmodem 300
Smartmodem 1200
Micromodem IIe

HUMAN ENGINEERED SOFTWARE (HESWARE)
HesModem I

IDEASSOCIATES INC.
IDEAComm 1200

LEXICON CORPORATION OF MIAMI
Lex-11
Lex-12
Lex-21

MFJ ENTERPRISES INC.
1237 Modem
1233 Modem

MICROBITS PERIPHERAL PRODUCTS INC.
MPP-1000

MORROW INC.
MM300

MULTI TECH SYSTEMS, INC.
MultiModem HC3
MultiModem
MultiModem HC
MultiModem PC

NEC AMERICA INC.
N212BR

NOVATION INC.
J-Cat
D Cat
103 Smart-Cat
103/212 Smart-Cat
212 Auto-Cat
Cat
Apple Cat II

PENRIL DATACOMM
8216
8216 Auto Dial

PROMETHEUS PRODUCTS INC.
ProModem 1200

RACAL-VADIC
VA212 LC
VA212 Auto Dial
VA3451 Auto Dial
Maxwell 300V
Maxwell 1200V
Maxwell 300PC
Maxwell 1200PC
VA1222

RIXON, INC.
R212A
PC212A

TEK-COM CORPORATION
P212A
P212C
Star 103

TELEVIDEO SYSTEMS
300/1200 Modem

THE MICROPERIPHERAL CORP.
The Microconnection

TNW CORPORATION
Operator 103

TRANSEND CORP.
PC Modemcard 1200
Transmodem 1200
Apple Modemcard

UNIVERSAL DATA SYSTEMS
108

103J
202S
202S/SS
212A
212A/D
212LP

U.S. ROBOTICS, INC.
Password
Password 300
Personal Communicator
S-100

VEN-TEL INC.
PC Modem Half-Card
PC Modem 1200
1200 Plus
HP 150 Internal Modem

VISIONARY ELECTRONICS
Visionary 100
Visionary 1200

ZOOM TELEPHONICS
Networker

COMMUNICATIONS SOFTWARE REVIEWS

ALPHA SOFTWARE CONNECTION
The Apple-IBM Connection

APPLE COMPUTER, INC.
Apple Access II

BI-TECH ENTERPRISES, INC.
Uniterm
Micro Link II Communications Package

COMMUNICATIONS RESEARCH GROUP, INC.
BLAST

COMPUTER: APPLICATIONS INC.
Apple-Link

DIGITAL MARKETING
The Micro Link II

DIRECT AID, INC.
Direct Connect
The Impersonator

DYNAMIC MICROPROCESSOR ASSOCIATES, INC.
Ascom

EXECS SOFTWARE
TermExec

HAWKEYE GRAPHIX
Commx-Pac

HAYES MICROCOMPUTER PRODUCTS, INC.
Smartcom I
Smartcom II

IDEASSOCIATES INC.
IDEACom 1200

LINDBERGH SYSTEMS, INC.
Omniterm
Omniterm 2

MARK OF THE UNICORN
PC/InterComm

MICROSTUF, INC.
Crosstalk

MICRO-SYSTEMS SOFTWARE
MicroTerm

MICROTECHNIC SOLUTIONS, INC.
Smart 64

MIDWEST MICRO INC.
SuperTerm

MORROW, INC.
ReachOut

MYCROFT LABS INC.
MITE

NEW ERA TECHNOLOGIES, INC.
Conexus
Mighty Mail
MIST

SANYO BUSINESS SYSTEMS
Sanycom

SOFTWARE PUBLISHING CORPORATION
PFS:Access

SOUTHEASTERN SOFTWARE
Data Capture PC
Data Capture IIe

SUPERSOFT INC.
Term II

TAU SYSTEMS
I/O Express

TRANSEND CORPORATION
Transend PC

UNITED SOFTWARE INDUSTRIES
ASCII Express
P-Term
Z-Term

U.S. ROBOTICS INC.
Copylink
Telpac

VECTOR GRAPHICS, INC.
Conect

WINDMILL SOFTWARE INC.
Video Link 88

ZOOM TELEPHONICS, INC.
Netmaster

Appendix B

Vendors Addresses and Phone Numbers

MODEM VENDOR LIST

APPLE COMPUTER, INC.
20525 Mariani Ave.
Cupertino, CA 95014
(408) 996-1010

ANCHOR AUTOMATION INC.
6913 Valjean Ave.
Van Nuys, CA 91406
(213) 997-6493

BUSINESS COMPUTER CORPORATION (BIZCOMP)
532 Weddell Dr.
Sunnyvale, CA 94089
(408) 733-7800

CERMETEK MICROELECTRONICS, INC.
P.O. Box 3565
1308 Borregas Ave.
Sunnyvale, CA 94088
(408) 752-5000

COMMODORE BUSINESS MACHINES INC.
1200 Wilson Drive
West Chester, PA 19380
(215) 431-9100

COMMUNICATIONS AND PERIPHERALS MARKETING CORP.
2318-B Walsh Ave.
Santa Clara, CA 95051
(408) 986-1555

COMPUTER COMMUNICATIONS SPECIALISTS, INC.
Atlanta/Peachtree Corners Business Park
6683 Jimmy Carter Blvd.
(404) 441-3114

CONCORD DATA SYSTEMS
303 Bear Hill Rd.
Waltham, MA 02154
(617) 890-1394

DATASOLVERS, INC.
509 S. Fern
Wichita, KS 67213
(316) 264-5068

GANDALF TECHNOLOGIES INC.
350 East Dundee Rd.
Wheeling, IL 60090
(312) 459-6630

GENERAL DATACOMM, INC.
One Kennedy Ave.
Danbury, CT 06810
(203) 797-0711

HAYES MICROCOMPUTER PRODUCTS, INC.
5923 Peachtree Industrial Blvd.
Norcross, GA 30092
(404) 441-1617

HUMAN ENGINEERED SOFTWARE (HESWARE)
150 North Hill
Brisbane, CA 94005
(415) 468-4110

IDEASSOCIATES INC.
7 Park Dr.
Bedford, MA 01730
(617) 275-4430

LEXICON CORPORATION OF MIAMI
1541 N.W. 65th Ave.
Fort Lauderdale, FL 33313
(800) 327-8913

MFJ ENTERPRISES, INC.
P.O. Box 494
Mississippi State, MS 39762
(800) 647-1800

MICROBITS PERIPHERAL PRODUCTS, INC.
225 W. Third St.
Albany, OR 97321
(503) 967-9075

MORROW, INC.
600 McCormick St.
San Leandro, CA 94577
(415) 430-1970

MULTI TECH SYSTEMS, INC.
82 Second Ave. S.E.
New Brighten, MN 55112
(612) 631-3550

NEC AMERICA, INC.
1012 Steward Dr.
Sunnyvale, CA 94086
(408) 737-7711

NOVATION INC.
18664 Oxnard St.
Tarzana, CA 91356
(213) 996-5060

PENRIL DATACOMM
207 Perry Parkway
Gaithersburg, MD 20877
(301) 921-8600

PROMETHEUS PRODUCTS, INC.
45277 Fremont Blvd.
Fremont, CA 94538
(415) 490-2370

RACAL-VADIC
222 Caspian Dr.
Sunnyvale, CA 94089-9874
(408) 744-0810

RIXON, INC.
2120 Industrial Parkway
Silver Spring, MD 20904
(301) 622-2121

TEK-COM CORPORATION
1887-C103 O'Toole Ave.
San Jose, CA 95131
(408) 263-7400

TELEVIDEO SYSTEMS, INC.
1170 Morse Ave.
Sunnyvale, CA 94086
(408) 745-7760

THE MICROPERIPHERAL CORP.
2565 152d Ave. N.E.
Redmond, WA 98052
(206) 881-7544

TNW CORPORATION
3444 Hancock St.
San Diego, CA 92110
(619) 296-2115

TRANSEND CORPORATION
2190 Paragon Dr.
San Jose, CA 95131
(408) 946-7400

UNIVERSAL DATA SYSTEMS
5000 Bradford Dr.
Huntsville, AL 35805
(205) 837-8100

U.S. ROBOTICS, INC.
1123 W. Washington Blvd.
Chicago, IL 60607
(312) 733-0497

VEN-TEL, INC.
2342 Walsh Ave.
Santa Clara, CA 95051
(408) 727-5721

VISIONARY ELECTRONICS
141 Parker Ave.
San Francisco, CA 94118
(415) 751-8811

SOFTWARE VENDOR LIST

ALPHA SOFTWARE CONNECTION
30 B St.
Burlington, MA 01803
(617) 229-2924

APPLE COMPUTER, INC.
20525 Mariani Ave.
Cupertino, CA 95014
(408) 996-1010

APPLIED COMPUTER TECHNIQUES
104 Knight Dr.
San Rafael, CA 94901
(415) 459-3212

BI-TECH ENTERPRISES, INC.
10 Carlough Rd.
Bohemia, NY 11716-2996
(516) 567-8155

COMMUNICATIONS RESEARCH GROUP, INC.
8939 Jefferson Highway
Baton Rouge, LA 70808
(504) 923-0888

COMPUTER: APPLICATIONS INC.
13300 SW 108 Street Circle
Miami, FL 33186
(305) 385-4277

DIGITAL MARKETING
2363 Boulevard Circle
Walnut Creek, CA 94595
(415) 947-1000

DIRECT AID, INC.
P.O. Box 4420
Boulder, CO 80306
(303) 442-8080

DYNAMIC MICROPROCESSOR
ASSOCIATES, INC.
545 Fifth Ave.
New York, NY 10017
(212) 986-5872

EXECS SOFTWARE
201 Waltham St.
Lexington, MA 02173
(617) 862-3170

HAWKEYE GRAPHIX
23914 Mobile St.
Canoga Park, CA 91307
(213) 348-7909

HAYES MICROCOMPUTER PRODUCTS, INC.
5923 Peachtree Industrial Blvd.
Norcross, GA 30092
(404) 441-1617

IDEASSOCIATES INC.
7 Oak Park Dr.
Bedford, MA 01730
(617) 275-4430

LINDBERGH SYSTEMS INC.
49 Beachmont St.
Worcester, MA 01609
(617) 852-0233

MARK OF THE UNICORN
222 Third St.
Cambridge, MA 02142
(617) 576-2760

MICROSTUF, INC.
1845 The Exchange
Atlanta, GA 30339
(404) 998-3998

MICRO-SYSTEMS SOFTWARE
4301-18 Oak Circle
Boca Raton, FL 33431
(305) 983-3390

MICROTECHNIC SOLUTIONS, INC.
P.O. Box 2940
New Haven, CT 06515
(203) 389-8383

MIDWEST MICRO INC.
311 West 72d St.
Kansas City, MO 64114
(816) 333-7200

MORROW, INC.
600 McCormick St.
San Leandro, CA 94577
(415) 430-1970

MYCROFT LABS, INC.
P.O. Box 6045
Tallahassee, FL 32314
(904) 385-1141

NEW ERA TECHNOLOGIES, INC.
2025 Eye St. N.W. Suite 924
Washington, DC 20006
(202) 887-5440

SANYO BUSINESS SYSTEMS
51 Joseph Street
Moonarchie, NJ 07074
(201) 440-9300

SOFTWARE PUBLISHING CORP.
1901 Landings Dr.
Mountain View, CA 94043
(415) 962-8910

SOUTHEASTERN SOFTWARE
6414 Derbyshire Dr.
New Orleans, LA 70126
(504) 246-8438

SUPERSOFT INC.
P.O. Box 1628
Champaign, IL 61820
(217) 359-2112

TAU SYSTEMS CORPORATION
7115 Leesburg Pike #211
Falls Church, VA 22043
(703) 237-7200

TRANSEND CORPORATION
2190 Paragon Dr.
San Jose, CA 95131
(408) 946-7400

UNITED SOFTWARE INDUSTRIES
1880 Century Park, Suite #311
Los Angeles, CA 90067
(213) 556-2211

U.S. DIGITAL CORPORATION
5699-D S.E. International Way
Milwaukie, OR 97222
(503) 654-0668

U.S. ROBOTICS, INC.
1123 W. Washington Blvd.
Chicago, IL 60607
(312) 733-0497

VECTOR GRAPHICS, INC.
500 North Ventu Park Rd.
Thousand Oaks, CA 91320
(805) 499-5831

WINDMILL SOFTWARE, INC.
2209 Leominister Dr.
Burlington, Ontario
Canada L7P 3W8
(416) 336-3353

ZOOM TELEPHONICS, INC.
207 South St.
Boston, MA 02111
(800) 631-3116

Glossary

acoustic coupler—A device used with a standard telephone to facilitate the transfer of data from a modem to the phone lines. It consists of two rubber cups—one for the earpiece and one for the mouthpiece of the handset of the phone.

analog loopback test—A test to verify the working condition of a modem and associated equipment. Data entered from the keyboard is passed to the modem and modulated, but instead of transferring the data to the phone line, it is fed back into the modem and then displayed on the screen.

answer mode—This term is used in two ways, depending on the type of modem. Modems transmitting a full-duplex use separate frequencies for transmitting. These frequencies are distinguished with the term originate or answer mode. Some intelligent modems use the term answer mode to describe the condition of being ready to answer an incoming call.

ASCII—An acronym for American Standard Code for Information Interchange. The code consists of 128 numbers ranging from 0 to 127. Each number represents an alphanumeric character, a punctuation mark, or a control code.

asynchronous—A method of serial transmission where characters are separated by special bits, called start and stop bits.

auto-answer—A feature that enables some modems to automatically answer an incoming call.

auto-dial—A feature that enables some modems to dial a number that has been entered through the keyboard of a terminal or computer.

baud—A measurement of transmission speed. It roughly equals the number of bits transmitted per second.

Bell 103A—A set of specifications provided by Bell Systems for modems transmitting at 0 to 300 baud.

Bell 212A—A set of specifications provided by Bell Systems for modems transmitting at 1200 baud. Bell 212A modems are also able to transmit using Bell 103A specifications.

Bell-compatible—Modems must use the same

specifications for transmitting data. Bell-compatible modems identify the Bell specifications to which a modem conforms.

bit—An acronym for "binary digit." Representing the smallest quantity of data processing information, seven or eight bits are grouped together to represent a single character.

bps.—Bits per second.

buffer—A portion of memory set aside as a temporary storage area for characters.

bus—A path, often involving several lines, over which electrical signals travel.

byte—A group of eight bits used to represent information within a digital device.

capture—A process used by communications software to temporarily save information as it is received through a modem.

carrier—A continuous frequency upon which other signals can be superimposed. A modem will look for this frequency to determine that a proper connection has been made with another modem.

central processing unit—That part of a computer which adds, subtracts, compares, and controls the movement of data within the computer.

character—Any alphabetic letter, punctuation mark, number, or special symbol.

chip—A term referring to an integrated circuit.

communications adapter—A device that disassembles bytes of data into individual bits for serial transmission. The same device will reassemble serial data back into bytes.

conditioning—A process of improving the signal quality of voice-grade line for transmission of data.

control characters—Special characters recognized by some devices that are used to control the flow and display of data.

conversation mode—A mode of operation where two operators on two systems can interactively converse using the keyboard of their terminals.

crossover cable—Also called a null modem, a cable for connecting and transferring data between the serial ports of two computers.

cyclical redundancy check—An error detection scheme involving the generation and verification of a polynomial value that depends on the contents of a message.

data compression—The process of reducing the amount of data by eliminating redundant characters and simplifying the representation of characters.

demodulator—The process of converting the frequencies used for transmitting data across phone lines into the digital signals used by computers.

digital loopback test—A test to verify the working condition of a modem and associated equipment. Data entered from a keyboard is passed to a modem, modulated, and transferred through the phone system to a receiving modem. The receiving modem, rather than pass the data to its computer, transmits it back to the original system.

download—The process of saving information as it is received through a modem. Usually the term is used to describe the process of saving directly to diskette.

duplex—See full-duplex or half-duplex.

EBCDIC—Extended Binary Coded Decimal Interchange Code. A code using eight bits to represent up to 256 characters developed for the IBM 360 computer.

echo—Often used in full-duplex operation, characters entered from the keyboard of one computer are received and returned by a second computer. The returning characters are displayed on the sender's screen and essentially provide a method of verifying that data has been transferred correctly.

electronic mail—An application of telecommunications where messages are sent by one computer and stored for later reading on a second computer.

error detection—The process of verifying that data transmitted by one system is received correctly by another.

escape character—A special character, often the ASCII value 27, that informs a device that the subsequent ASCII values will have new meaning.

flow control—A procedure enabling a terminal receiving data to command the transmitting device to stop or start transmitting.

frequency-shift keying—A method of transmitting information by representing the one digital value with one frequency and the zero digital value with a second frequency.

full-duplex—A communications procedure allowing information to be transmitted and received simultaneously.

function keys—Special keys found on some keyboards that can be programmed to transmit assigned values.

half-duplex—A communications procedure where information is either transmitted or received, but not both simultaneously.

handshaking—An exchange of signals between two devices communicating to each other when each is ready to transmit or receive data.

hardware—Mechanical, electrical, or magnetic devices comprising the physical components of a system.

hardwired—A physically wired connection between two devices.

hexadecimal—A numbering system based on 16. Values 0 through 9 are represented with numerals, and values 10 through 15 are represented by characters A through F.

host—The primary or controlling computer when two or more computers are communicating with each other. Usually when a personal computer accesses a commercial system, the mainframe computer is considered the host.

intelligent—A descriptive term applied to digital devices that contain a microprocessor. Devices with microprocessors generally contain more sophisticated features than those without.

interface—An intermediate device between two other devices that provides the necessary components to ensure compatibility.

interrupt—A signal that instructs a microprocessor to temporarily halt its current operation to perform another.

logoff—A formal procedure of disconnecting with a host computer. It generally involves transmitting special commands recognized by the host.

logon—A formal procedure of connecting with a host computer. It generally involves answering a series of questions transmitted by the host.

macro—A feature found on some programs that enables a user to assign special meaning to a key or command. Entering that key or command causes a series of programmed characters or commands to be transmitted automatically.

mainframe—Used loosely to describe a large, sophisticated computer capable of accepting and processing information from a series of terminals simultaneously.

mark—A condition signifying the presence of a binary one.

menu—A table of options displayed on the screen.

microcomputer—A class of computers associated with personal and home computers and designed to perform one task at a time.

microprocessor—A device that processes data in response to programmed instructions.

minicomputer—A class of computers smaller than mainframe computers but larger than microcomputers.

mode—A condition within a device or system for performing a specified set of features. The same command may perform different tasks in different modes.

modem—A device for converting the signals transmitted by digital devices into signals that can be transferred across a phone line and vice versa.

network—An arrangement where a number of computers and peripherals are linked together to share resources.

null modem cable—Also called a crossover cable, a cable for connecting and transferring data between the serial ports of two computers.

off line—A state where two communications devices are not physically connected.

on line—A state where two communications devices are physically connected.

originate mode—Modems transmitting in full duplex use separate frequencies for transmitting. These frequencies are distinguished with the term originate or answer mode.

parallel—The process of transmitting all bits in a byte simultaneously.

parameters—Limits placed on a system when transferring information. Used often to describe limits that must be adhered to in RS-232C transmissions, including baud rate, parity, duplex, word length, and start and stop bits.

parity—A method of error detection where the eight bit of an ASCII byte is set to make the sum total of all bits in the byte even. Odd parity sets the eighth bit to make the sum odd. A receiving device adds the bits and confirms the parity. If the sum is not even in even parity, then the receiving device requests the byte to be transmitted again.

peripheral—Any device connected to a computer and dependent on it for instructions.

phase shifting—A process used in telecommunications where information is transferred by altering the phase of a carrier frequency.

program—A set of instructions that controls the activities of a computer.

protocol—A defined procedure of transferring information.

RAM—Random Access Memory; memory devices that can be read, and altered by the user. Information on memory is lost when power is removed from the device.

RJ11C—A standardized jack used on modular telephone and modems.

ROM—Read-Only Memory; memory devices that can be read but not altered by the user. Information is not lost when power is removed.

RS-232C—Recommended Standard 323 version C. A standard for serial transmission of data developed by the Electronics Industry Association.

scroll—The movement of text on the screen as new information is received.

semiconductors—A term used to describe the fundamental building blocks of digital systems.

serial interface—A device that disassembles bytes of data into individual bits for serial transmission. The same device will reassemble serial data back into bytes.

serial transmission—Transmission of data one bit at a time.

software—Programs which provide instructions to a computer to perform a designated function.

spooling—The process of storing data in a special buffer to allow a receiving device to receive it at its own speed.

start bit—A special bit longer in duration than normal data bits that indicates the beginning of a new byte of data in asynchronous transmission.

stop bit—A special bit longer in duration than normal data bits that indicates the end of a byte of data in asynchronous transmission.

synchronous transmission—A method of serial transmission where control characters indicate the beginning and end of a stream or block of data.

terminal—Any device designed for sending or receiving information from a computer. The device generally includes a keyboard and screen. Intelligent terminals include features beyond simply sending or receiving.

terminal emulator—A program that configures a computer to behave like a specified terminal.

transistor—A device that allows a voltage source to regulate a resistance value. Microscopic transistor are the fundamental unit of integrated circuits.

upload—A process accomplished through communications software where a file is transferred from a diskette through a modem to a receiving system.

VA3400—A standard developed by Racal-Vadic for full-duplex 1200-baud transmission.

XMODEM protocol—A procedure existing in public domain for transferring data serially with complete error detection.

XON/XOFF protocol—A procedure involving flow control codes for transferring data between two systems. Both systems can tell the other when to start or stop transmitting.

Modem Comparison Chart

Manufacturer Model	Special Features	Baud Rate	Protocol	Type	Interface	Price	Software Included
Anchor Automation Volksmodem		300	Bell 103	E	RS-232C	$ 79.95	No
Anchor Automation Signalman Mark I		300	Bell 103	E	RS-232C	99.00	No
Anchor Automation Signalman Mark III	exclusive for TI-99/4A	300	Bell 103	E	RS-232C	139.00	No
Anchor Automation Signalman Mark X	auto answer/auto dial	300	Bell 103 113	E	RS-232C	169.00	No
Anchor Automation Signalman XII	auto answer/dial/redial	300/ 1200	Bell 212A/ 103/113	E	RS-232C	399.00	No
Apple Computer Modem 300	auto answer/dial	300	Bell 103	E	RS-232C	225.00	Yes (Apple Term)
Apple Computer Modem 1200	auto answer/dial	300/1200	Bell 212/ 103	E	RS-232C	495.00	Yes (Apple Term)
Bizcomp 1022 Intelligent	auto dial/answer; command - driven	300	Bell 103/ 113	E	RS-232C	279.00	No
Bizcomp 1012 Intelligent	auto dial/answer	1200/300	Bell 212A/ 103/113	E	RS-232C	549.00	No
Bizcomp PC:IntelliModem	IBM PC/XT computers auto answer/dial; dialing directory	300/1200	Bell 212	I	IBM	499.00	Yes (PC:Intellicom)
Cermetek Info-Mate 212A	auto answer/dial; 52-location memory	300/1200	Bell 212A/ 103	E	RS-232C	595.00	No
Cermetek Info-Mate 212PC	IBM PC/XT, compatibles; auto answer/dial; 52-location memory	300/1200	Bell 212A/ 103	I	IBM	480.00	Yes (Modem-Mate)
Commodore VICModem	Commodore 64, VIC 20 only	300	Bell 103	I	Commodore	69.95	Yes
Commodore Automodem	Commodore 64, VIC 20 only; auto answer/dial	300	Bell 103	I	None	149.95	Yes (Term 64)
Communications & Peripherals MM-102		300	Bell 103	E	RS-232C	64.95	No
Communications & Peripherals AutioModem 1200A	auto answer/dial	300/1200	Bell 103/ 212A	E	RS-232C	279.95	No
Computer Communications Specialists AudioModem	100 words and phrases can be stored; auto answer/dial	1200	202S	E	RS-232C	4895.00	Yes
Concord Data 224 Data Modem	auto answer/redial	2400/1200	Bell 212	E	RS-232C or CCITTV.24/V.28	995.00	No
Concord Data 224 Autodial Modem	auto answer/dial	2400/1200	Bell 212	E	RS-232C or CCITV.24/V.28	1195.00	No

Manufacturer Model	Special Features	Baud Rate	Protocol	Type	Interface	Price	Software Included
Datasolvers, Inc. DS1200SMT	Kaypro computers only; Hayes-compatible; auto answer/dial	1200/300	Bell 212A	I	Kaypro	427.50	Yes (SMODEMK)
Gandalf Data SAM 212A	52-location memory; auto answer/dial	300/1200	Bell 103/ 212A	E	RS-232C	618.00	No
General Datacomm 201C		2400/1200	Bell 201C/ CCITT V.26	E	RS-232C	975.00	No
General Datacomm 212A/L	brown-out protection; auto answer	1200/300	Bell 103/ 113/212A	E	RS-232C	675.00	No
General Datacomm 212A/ED	auto answer/ dial/redial	1200/330	Bell 103/ 113/212A	E	RS-232C	575.00	No
Hayes Microcomputer Smartmodem 300	auto answer/dial	300	Bell 103	E	RS-232C	289.00	No
Hayes Microcomputer Smartmodem 1200	auto answer/dial	1200/300	Bell 212A/ 103	E	RS-232C	699.00	No
Hayes Microcomputer Micromodem IIe	Apple computers only; auto answer/dial; menu	300	Bell 103	I	Apple	329.00	Yes (Smartcom I)
Hesware Hesmodem I	Commodore 64, VIC 20 only; 2 free hours on CompuServe	300	Bell 103	I	Commodore	69.95	Yes
IDEAssociates IDEAComm 1200	IBM PC/XT; auto answer/dial; menu-driven software	1200/300	Bell 212A/ 103	I	IBM	495.00	Yes (IDEAComm 1200)
Lexicon Corp. Lex-11	acoustic-coupled; lightweight	300	Bell 103A	E	RS-232C	145.00	No
Lexicon Corp. Lex-12	acoustic-coupled	300	Bell 103A	E	RS-232C	165.00	No
Lexicon Corp. Lex-21	built into a portable terminal and printer	300	Bell 103A	I	---	395.00	Yes
MFJ Enterprises 1237 Modem	Commodore 64, VIC 20 only	300	Bell 103	I	Commodore	49.95	Printed Listing
MFJ Enterprises 1233 Modem		300	Bell 103	E	RS-232C	129.95	No
Microbits Peripheral Products MPP-1000	ATARI; print feature; auto dial; 9 buffers can be created; menu-driven	300	Bell 103	E	none	149.95	Yes
Morrow MM300	Micro Decision computers; auto answer/ dial; dialing directory can be created	300	Bell 103/ 113	E	Micro Dec.	299.00	Yes (ReachOut)
Multi Tech Systems MultiModem PC	IBM PC/XT, compatibles; Hayes-compatible; auto answer/dial; menu of 20 phone numbers	1200/300	Bell 212A/ 103/113	I	IBM	549.00	Yes (MultiCom)

Manufacturer Model	Special Features	Baud Rate	Protocol	Type	Interface	Price	Software Included
Multi Tech Systems MultiModem HC3	Hayes-compatible; auto answer/dial/redial	300	Bell 103	E	RS-232C	289.00	No
Multi Tech Systems MultiModem	Hayes-compatible; auto answer/dial/redial; memory for 6 numbers	1200/300	Bell 212/ 103	E	RS-232C	$ 549.00	No
Multi Tech Systems MultiModem HC	Hayes-compatible; auto answer/dial/redial	1200/300	Bell 212; 103	E	RS-232C	549.00	No
NEC America N212BR	auto answer/dial; memory for 12 numbers and messages	1200/300	Bell 212A	E	RS-232C	795.00	No
Novation 103 Smart-Cat	auto answer/dial	300	Bell 103	E	RS-232C	249.00	No
Novation 103/212 Smart-Cat	auto answer/dial	1200/300	Bell 212A/ 103/113	E	RS-232C	595.00	No
Novation D Cat		300	Bell 103/ 113	E	RS-232C	199.00	No
Novation J-Cat	auto answer/dial	300	Bell 103/ 113	E	RS-232C	149.00	No
Novation Cat	acoustic-coupled	300	Bell 103/ 113	E	RS-232C	189.00	No
Novation 212 Auto-Cat	auto answer/dial	1200/300	Bell 212A/ 103/113	E	RS-232C	695.00	No
Novation Apple Cat II	Apple II, II+, requires 48K memory; 27K buffer; auto answer/ dial	1200/300	Bell 103/ 202/212 w/upgrade	I	Apple	595.00 w/upgrade	Yes (Com Ware II)
Penril Datacomm 8216 Auto Data 300/1200	auto answer	300/1200	Bell 103/ 113/212A	E	RS-232C	525.00	No
Penril Datacomm 8216 w/auto dial	up to 10 stored numbers; auto answer/ dial	300/1200	Bell 103/ 113/212A	E	RS-232C	650.00	No
Prometheus Products ProModem 1200	auto answer/dial; Hayes-compatible; built-in clock, calendar	1200/300	Bell 212A/ 103/113	E	RS-232C	495.00	No
Racal Vadic Maxwell 300PC	IBM PC/XT, portable, compatible's, Hayes-com- patible; auto answer/ dial; tandem dialing	300	Bell 103/ 113	I	IBM	325.00	Yes (George)
Racal Vadic Maxwell 1200PC	IBM PC/XT/portable, compatibles; tandem dialing; auto dial/ answer; Hayes - compatible	1200/300	Bell 212A/ 103/113	I	IBM	595.00	Yes (George)
Racal Vadic Maxwell 300V	auto answer/dial; Hayes- compatible; tandem dialing	300	Bell 103/ 113	E	RS-232C	350.00	No

Manufacturer Model	Special Features	Baud Rate	Protocol	Type	Interface	Price	Software Included
Racal Vadic Maxwell 1200V	auto dial/answer; Hayes-compatible; tandem dialing	1200/300	Bell 212A/ 103/113	E	RS-232C	$595.00	No
Racal Vadic VA1222	auto answer	1200/600/ 300	V.24, V.22; V.25, V.28, V.54	E	RS-232C V.24/V.28	975.00	No
Racal Vadic Auto Dial VA212	auto answer/dial/ redial; command-driven	1200/300	Bell 103/212	E	RS-232C	695.00	No
Racal Vadic VA212 LC	auto answer; auto self-text	1200/300	Bell 103/212	E	RS-232C	495.00	No
Racal Vadic VA 3451 Auto Dial	auto answer/dial/ redial; command-driven	1200/300	Bell 103/ 212; VA 3440	E	RS-232C	875.00	No
Rixon PC212A	IBM PC; blind dial; auto answer/dial; Hayes-compatible	1200/300	Bell 212A 103/113	I	IBM	539.00	Yes (PC Com I)
Rixon R212A	auto answer/dial; Hayes-compatible; stores up to 10 numbers, 60 characters each	1200/300	Bell 212A 103/113	E	RS-232C	899.00	No
Tek-Com P212A	auto answer; programmed, fixed loop, permissive phone lines	1200/300	Bell 212A	E	RS-232C	595.00	No
Tek-Com P212C	auto answer; programmed, fixed loop, permissive phone lines	1200/300	Bell 212A	E	RS-232C	650.00	No
Tek-Com Star 103	acoustic coupled; command-driven	300	Bell 103/ CCITT V.21	E	RS-232C	199.00	No
Televideo Systems Personal Terminal	Televideo Personal Terminal; auto answer/ dial; command-driven	1200/300	Bell 212A	----	----	549.00	Yes
The Microperipheral Corp. The Microconnection	ATARI, C-64, VIC 20; auto answer/dial	300	Bell 212A	E	---	179.95	Program listing
TNW Operator 103	auto answer/dial; fast dial; command-driven	300	Bell 103	E	RS-232C	169.00	No
Transend PC Modemcard 1200	IBM PC/XT; subscription to SOURCE; auto answer/ dial; Hayes-compatible; command-driven	1200/300	Bell 212A/ 103	I	IBM	549.00	Yes
Transend TransModem 1200	auto answer/dial/ redial; blind dial	1200/300	Bell 212A/103	E	RS-232C	695.00	No
Transend Apple Modemcard	Apple II, //e; subscription to SOURCE; auto answer/dial	110/300	Bell 103	I	Apple	275.00	No
Universal Data Systems 108	private line connection only	300	Bell 103	E	RS-232C	295.00	No

Manufacturer Model	Special Features	Baud Rate	Protocol	Type	Interface	Price	Software Included
Universal Data Systems 103J	auto answer; various phone connections by request	300	Bell 103/ 113	E	RS-232C	$425.00	No
Universal Data Systems 202S	auto answer	0-1200	Bell 202	E	RS-232C	475.00	No
Universal Data Systems 202S/SS	auto answer	0-1200	Bell 202	E	RS-232C	550.00	No
Universal Data Systems 212A	auto answer; command-driven	0-300/ 1200	Bell 103/ 113/212A	E	RS-232C	595.00	No
Universal Data Systems 212A/D	auto answer/dial; 5 numbers in memory w/up to 30 digit or command characters	0-300/ 1200	Bell 103/ 113/212A	E	RS-232C	645.00	No
Universal Data Systems 212LP		1200	Bell 212	E	RS-232C	345.00	No
U.S. Robotics Password	auto answer/dial	1200/300	Bell 212A/103	E	RS-232C	449.00	No
U.S. Robotics Password 300	auto answer/dial	300	Bell 103	E	RS-232C	199.00	No
U.S. Robotics Personal Communicator	IBM PC/XT/compatibles; auto answer/dial	1200/300	Bell 212A/103	I	IBM	499.00	Yes (Telpac)
U.S. Robotics S-100	S-100 computers; auto answer/dial; series of commands, 60 chars in length	1200/300	Bell 212A/103	I	S-100	449.00	No
Ven-Tel PC Modem Half-Card	IBM PC/XT, compatibles; Compaq; auto answer/ dial; Hayes-compatible	1200/300	Bell 212A/ 103	I	IBM	549.00	Yes (Crosstalk XVI)
Ven-Tel PC Modem 1200	IBM PC/XT, compatibles, Compaq; auto answer/ dial; Hayes-compatible	1200/300	Bell 212A/ 103	I	IBM	499.00	Yes (Crosstalk XVI)
Ven-Tel 1200 Plus	auto answer/dial; command sets	1200/300	Bell 212A/103	E	RS-232C	495.00	No
Ven-Tel HP 150 Internal	Hewlett Packard-compatible; auto answer/ dial	1200/300	Bell 212A/103	I	HP150	425.00	No
Visionary Electronics Visionary 100	auto answer/dial; 2K memory, expands to 48K; suitable for electronic mail	300	Bell 103	E	RS-232C	595.00	No
Visionary Electronics Visionary 1200	auto answer/dial; 2K memory, expandable to 48K; suitable for electronic mail	1200/300	Bell 212A/103	E	RS-232C	795.00	No

*I=internal; E=external

Index

A

Acoustic connections, 27
Alpha Software Connection, 129-130
  The Apple-IBM Connection, 129-130
Analog loopback text, 16
Anchor Automation, Inc., 37-42
  Signalman Mark I, 38-39
  Signalman Mark III, 39
  Signalman Mark X, 39-40
  Signalman Mark XII, 40-42
  Volksmodem, 37-38
Answering, 31
Answer mode, 14
Apple Computer, Inc., 42-43, 130
  Apple Access II, 130
  Apple Modem 300, 42-43
  Apple Modem 1200, 43
ASCII, 10
Asynchronous transmission, 15
Auto answer, 16
Auto dial, 15-16

B

Batch, 149
Baud rate, 24
Bell standards, 18-19
  Bell 103, 18
  Bell 113, 18
  Bell 202, 18
  Bell 212A, 18-19
Binary logic, 8
Binary numbering, 8
Binary point, 8
Binary system, 7
Bit, 9
Bi-Tech Enterprises, Inc., 130-131
  Micro Link, 130-131
  Uniterm, 131
Bizcomp, 43-47
  1012 Intelligent Modem, 45
  1022 Intelligent Modem, 44-45
  PC:IntelliModem, 46-47
Block of data, 15
BRS After Dark, 1
Buffer, 31
Bundled package, 34
Bus lines, 10
Byte, 9

C

Capturing data, 31
Carrier-detect LED, 16
Carrier detect line, 23
CCITT standards, 20
Central Processing Unit, 8
Cermetek Microelectronics, Inc., 47-48
  Info-Mate 212A, 47
  Info-Mate 212PC, 48
Character, 9
Chips, 7
Clear to send signal, 23
Command state, 28
Commodore Business Machines Inc. 48-50
  Automodem 1650, 49-50
  VICMODEM 1600, 48-49
Communications adaptor, 13
Communications and Peripherals Marketing Corp., 50-51
  Automodem 1200A, 50-51
  MM-102, 50
Communications program, vii
Communications Research Group, Inc., 131-132
  BLAST, 132
Communications software, 17, 28-32
  control of modem, 28-29
  main features, 30-32
  program drive, 29-30
Communications software features, 30-32
  answering, 31
  configuring, 30
  dialing, 30-31
  transferring data, 31-32
Communications software selection, 34-36
  applications, 35-36

compatibility, 34-35
reviews, 36
CompuServe, 2
Computer: Applications, Inc., 132-133
Apple-Link, 132-133
Computer Communications Specialists, Inc., 51-52
Audiomodem, 51-52
Concord Data Systems, 52-55
224 Autodial Modem, 43-55
224 DATA Modem, 53
Configuring, 30

D

Data communications equipment, 22
Data set ready line, 23
Data signal rate selectors, 23
Datasolvers, Inc., 55-56
DS1200SMT, 55-56
Data terminal equipment, 22
Data terminal ready line, 23
Data transmission, 10-11
information interchange codes, 10-11
signal codes, 10
Decem, 7
Decimal system, 10
Demodulation, 12
Dialing, 30-31
Dialog, 2-3
Digital loopback test, 16
Digital Marketing, 133
Micro Link II, 133
Digits, counting, 6-7
Digits and data, 6-11
binary numbering, 8
combinations, 7-8
counting digits, 6-7
microprocessors, 8-9
programming, 9-11
semiconductors, 7
Digitus, 7
Direct Aid, Inc., 133-135
Direct Connect, 133-134
The Impersonator, 134-135
Dow Jones News/Retrieval Service, 3
Downloading, 31
Duplex, 14, 24
Dynamic Microprocessor Associates, Inc., 135-136
ASCOM, 135-136

E

EBCDIC, 11
Electronic catalogs, 5
Electronic extension classes, 5
Electronic mail, 4-5
bulletin boards, 4-5
service, 4
Execs Software, 136
Term Exec, 136
External modem, 16

F

Firmware, 9
Frequency shifting, 13-14
Full duplex, 14

G

Gandalf Data, Inc., 56-57
SAM 212A, 56-57
General Datacomm, Inc., 57-61
201C, 57-59
212A/ED, 60-61
212A/L, 59-60

H

Half duplex, 14
Hawkeye Graphix, 136-137
Commx-pac, 137
Hayes Microcomputer Products, Inc. 61-65, 137-138
Micromodem IIE, 64-65
Smartcom I, 137-138
Smartcom II, 138
Smartmodem 300, 62-63
Smartmodem 1200, 63-64
Hesware, 65-66
Hesmodem I, 65-66
Homebanking, 5

I

IDEAssociates, Inc., 66-67, 138-139
IDEAComm 1200, 138-139
IDEAComm 1200 modem, 66-67
Information interchange codes, 10-11
Integrated circuits, 7
Intelligent modems, 15-16
auto answer, 16
auto dial, 15-16
internal speaker, 16
LED displays, 16
self test, 16
Internal modem, 16
Internal speaker, 16

L

LED displays, 16
Lexicon Corporation of Miami, 67-71
Lex-11, 68-69
Lex-12, 69
Lex-21, 69-71
Light-emitting diodes (LED), 16
Lindbergh Systems, Inc., 139-141
Omniterm, 139-140
Omniterm, 140-141

M

Machine language, 9
Mark, 13
Mark of the Unicorn, 141-142
PC/InterComm, 141-142
Memory, 9-10
MFJ Enterprises, Inc., 71-72
1233 Modem, 71-82
1237 Modem, 71
Microbits Peripheral Products, Inc., 72-73
MPP-1000C, 72-73
Microprocessors, 8-9
Microstuf, Inc., 142-143
Crosstalk, 142-143
Micro-Systems Software, 143-144
Micro Term, 143-144
Microtechnic Solutions Inc., 144
Smart 64, 144
Midwest Micro Inc., 144-145
SuperTerm, 144-145
Modem, vii, 12-17
communications software, 17
intelligent modems, 15-16
modem arrangements, 16-17
serial transmission 12-15
Modem arrangements, 16-17
Modem control, 28-29
Modem selection, 33-34
applications, 33
communications software, 34
computer compatibility, 33-34
reviews, 34
Modem standards, 18-24
Bell standards, 18-19
beyond the standards, 24
CCITT standards, 20
other interface standards, 24
RS-232C interface, 21-24
Vadic 3400, 19-20
Modular connections, 26-27
Modulation, 12
Morrow, Inc., 73-75, 145-146
MM 300, 74-75
ReachOut, 145-146
Multi Tech Systems, Inc., 75-81
MultiModem, 78-79
MultiModem HC, 79-81
MultiModem HC3, 77-78
MultiModem PC, 75-77
Mycroft Labs, Inc., 146-147
MITE, MITE/MS, or MITE 86, 146-147

N

NEC America, Inc., 81
N212BR, 81
Networking, 5
New Era Technologies, Inc., 147-149
Conexus, 147
Mighty Mail, 147-148
MIST/MIST+, 148-149
Novation Inc., 81-88
Apple Cat II, 87-88
Cat, 84-85
D Cat, 83
J Cat, 83-84
103 Smart-Cat, 81-82
103/212 Smart-Cat, 82-83
212 Auto-Cat, 85-87

Null modem, 22
Nybble, 9

O

On-line state, 28
Originate mode, 14

P

Packet-switched networks, 25-26
Parallel, transfer, 10
Parity, 24
Penril Datacomm, 88-91
  8216, 88-90
  8216 with auto dial, 90-91
Permissive devices, 26
Phase shifting, 13, 14
Phone connection, 25-27
  acoustic connections, 27
  modular connections, 26-27
  packet-switched networks, 25-26
  private lines, 25
  switched network, 25
Positional notation, 8
Potential, 8
Power-on LED, 16
Print mode, 29
Private lines, 25
Processing, 8
Program, 9
Program drive, 29-30
Programming, 9-11
  data transmission, 10-11
  memory, 9-10
Prometheus Products, Inc. 91-93
  ProModem 1200, 91-93

R

Random access memory, 9
Read-only memory, 10
Receive clock line, 23
Request to send signal, 23
Ring indicator line, 23
Rixon, Inc., 101-104
  PC212A, 102-103
  R212A, 103-104
RJ11C connection, 26
RJ41S connection, 27
RJ45S connection, 26
Routines, 148
RS-232C control signals, 22-23
  carrier detect, 23
  clear to send, 23
  data set ready, 23
  data signal rate selectors, 23
  data terminal ready, 23
  request to send, 23
  ring indicator, 23
  signal quality detector, 23
RS-232C interface, 21-24
  control signals, 22-23
  secondary channel, 24
  terminal and communications equipment, 22
  timing signals, 23-24
RS-232C secondary channel, 24
RS-422 standard, 24
RS-423 standard, 24
RS-449 standard, 24

S

Sanyo Business Systems, 149
  Sanycom, 149
Scripts, 145
Self test, 16
Semiconductors, 7
Serial card, 13
Serial transfer, 10
Serial transmission, 12-15
  duplex, 14
  frequency shifting, 13-14
  phase shifting, 14
  synchronization, 14-15
Signal codes, 10
Signal quality detector, 23
Software, 9
Software Publishing Corporation, 149-150
  PFS:Access, 150
Solid-state, 7
SOURCE, 3-4
Southeastern Software, 150-151
  Data Capture PC, 150-151
  Data Capture IIe, 151
Space, 13
Start and stop bits, 24
Supersoft Inc., 151-152
  Term II, 151-152
Switch, 7
Switched network, 25
Synchronization, 14-15
Synchronous transmission, 14

T

Tek-Com Corporation, 104-106
  P212A, 104-105
  P212C, 105-106
  Star 103, 106
Televideo Systems, Inc., 106-107
  Personal Terminal, 106-107
Terminal and communications equipment, 22
Terminal emulation, 30
The Microperipheral Corporation, 107-108
  The Micronnection, 107-108
Throttling data, 32
Timing signals, 23-24
TNW Corporation, 108
  Operator 103, 108
Transend Corporation, 109-112, 153
  Apple Modemcard, 111-112
  I/O Express, 152-153
  PC Modemcard 1200, 109-110
  Tau Systems Corporation, 152-153
  Transend PC, 153
  Transmodem 1200, 110-111
Transfer mode, 29
Transferring data, 31-32
Transistors, 7
Translation tables, 11
Transmit-clock lines, 23

U

United Software Industries, 153-156
  ASCII Express, 153-154
  P-Term, 154-155
  Z-Term, 155-156
Universal Data Systems, 112-118
  103J, 113-114
  108, 112-113
  202S, 114
  202S/SS, 114-115
  212A, 115-116
  212A/D, 116-117
  212 LP, 117-118
Uploading data, 32
U.S. Digital Corporation, 156-157
  Copylink, 156-157
U.S. Robotics, Inc., 118-121, 157
  Password, 118-119
  Password 300, 119
  Personal Communicator, 119-120
  S-100, 120-121
  Telpak, 157

V

Vadic 3400, 19-20
Vector Graphics, Inc., 157-158
  Conect, 158
Ven-Tel, Inc., 121-127
  HP 150 Internal Modem, 125-127
  1200 Plus, 124-125
  PC Modem Half-Card, 121-122
  PC Modem 1200, 122-124
Visionary Electronics, 127-128
  Visionary 100, 127
  Visionary 1200, 128
Volatile memory, 9

W

Windmill Software, Inc., 158-159
  Video Link 88, 158-159
Word, 9
Word length, 24

Z

Zoom Telephonics, Inc., 159
  Netmaster, 159

Edited by Suzanne L. Cheatle